China's Electricity Sector

Leo Lester • Mike Thomas
Editors

China's Electricity Sector

THE
LANTAU
GROUP

palgrave
macmillan

Editors
Leo Lester
The Lantau Group
Kwai Fong, Hong Kong

Mike Thomas
The Lantau Group
Kwai Fong, Hong Kong

ISBN 978-981-10-8191-0   ISBN 978-981-10-8192-7   (eBook)
https://doi.org/10.1007/978-981-10-8192-7

Library of Congress Control Number: 2018936660

Cover illustration: Mono Circles © John Rawsterne/patternhead.com

Printed on acid-free paper

This Palgrave Pivot imprint is published by Springer Nature
The registered company is Springer Nature Singapore Pte Ltd.
The registered company address is: 152 Beach Road, #21-01/04 Gateway East, Singapore 189721, Singapore

CONTENTS

List of Contributors

Philip Andrews-Speed is a senior principal fellow at the Energy Studies Institute, National University of Singapore. He has 35 years of experience in the field of energy and resources, starting his career as a mineral and oil exploration geologist before moving into the field of energy and resource governance. Until 2010 he was Professor of Energy Policy at the University of Dundee and Director of the Centre of Energy, Petroleum and Mineral Law and Policy. His main research interest is the political economy of energy and resource governance. His recent books include *The Governance of Energy in China: Transition to a Low-Carbon Economy.*

Huadong Dai is a Master's research student at the Southwestern University of Finance and Economics. His research interests include electricity market and financing issues. Prior to joining the university, he worked as an engineer and then marketing associate in the Wind Power Department, Dongfang Electric Corporation between 2010 and 2015. During this period, he was involved in multiple equity mergers and acquisitions and financial leasing projects, and was responsible for market, finance and policy analysis.

Xinmin Hu is a principal at The Lantau Group. He has a strong background in the Chinese power sector as well as extensive experience in international electricity and carbon markets, having been based in Australia for a number of years, most recently as an associate director with ReputTex (AU) and before that as an energy sector consultant. He is a regular reviewer for several international energy, operations research and optimization journals. He holds a PhD in operations research with a minor in economics from the University of Melbourne and an Msc in applied mathematics from Jilin University of Technology, China. He is fluent in both English and Mandarin.

Xiying Liu is an energy economist by training and holds a PhD degree in energy economics. She has worked on dozens of projects in the energy field, with a focus on the power sector, particularly in China, Singapore, other ASEAN member states and Germany. She advises decision makers from government and industry on energy markets and policies. She started her career as an assistant professor in the School of Energy Research at Xiamen University, China, before joining the Energy Studies Institute at the National University of Singapore, and most recently is working as an energy economist at an environmental fintech company in Switzerland.

Ying-Zi Wang is an energy and environmental economics Master's student at the Institute of Science and Development, Chinese Academy of Sciences. Her recent research has focused on the design and evaluation of energy and environmental policy, particularly the establishment of carbon markets and low carbon development paths for both traffic and the power sector in China. Past projects include Research on Energy Saving and Reform of Energy Consumption Pattern in Transportation, Planning on Response to Climate Change in Qinghai Province, and Research on Behavioral Characteristics and Mechanism Differences of Carbon Emission Trading Markets.

Dayong Zhang received his PhD from the University of Birmingham, UK, in 2006 and is Professor of Economics at Southwestern University of Finance and Economics, China. His research covers broad areas of economic and financial issues in emerging economies. His recent research interest specializes in energy economics, energy finance, banking and financial market in China. He has published over 30 peer-reviewed articles and he is an editorial board member of Emerging Market Finance and Trade and served on the board of directors of the Society for the Study of Emerging Markets.

Sufang Zhang is a professor at the School of Economics and Management, North China Electric Power University (NCEPU), Director of the Institute for Low Carbon Economics and Trade affiliated to the NCEPU, and Honorary Associate of International Development Studies based at the University of Sussex, UK. She had been a visiting scholar at University of Washington, USA, and Carleton University, Canada. Her research field focuses on China's renewable energy policy. She has directed or participated in many energy research projects, and many of her peer-reviewed papers have been published in renowned academic journals such as *Energy Policy*, *Applied Energy*, and *Energy*.

Lele Zou is an associate professor at the Institute of Science and Development, Chinese Academy of Sciences. Her research focuses on the modelling, analysis and assessment of the energy-environment-economic system, including policy analysis, and the modelling of energy and related sectors; the impacts of new low carbon technologies and markets; and vulnerability to environmental change at both global and regional scales. Her research uses methods and tools from economics and other social sciences combined with engineering. More than 40 peer-reviewed international journal papers and book chapters of hers have been published, and she has chaired more than 20 national-level research projects.

ABOUT THE EDITORS

Leo Lester has over a decade of international energy experience, having worked in the UK, India, Thailand and Saudi Arabia, first with a multinational oil and gas company in upstream business analysis and strategic planning, and later with the King Abdullah Petroleum Studies and Research Centre's work in North East Asia. He is a principal at The Lantau Group. He holds degrees from the Universities of Oxford (MA) and Reading (Phd), and is a global research fellow at the Institute of Asia and Pacific Studies. He is a CFA charter holder and a certified Financial Risk Manager.

Mike Thomas is a partner at The Lantau Group and has advised energy sector stakeholders on sensitive regulatory, commercial and strategic matters for nearly 30 years. He is an expert in the rigorous analysis of energy sector decisions including how or whether to regulate; how and when to rely on market forces; and the evaluation of opportunities and risks. Prior to co-founding The Lantau Group, he headed the Asia Pacific Energy & Environment practice of a global consulting firm. He has been based in the Asia Pacific region since early 1997. He has an MPP from Harvard Kennedy School and a BA in economics from Carleton College.

Abbreviations

ABS   Asset-Backed Security
CASISD   Institute of Science and Development at the Chinese Academy of Sciences
CEPLM   Coal-Electricity Price Linkage Mechanism
CPC   Communist Party of China
CSG   China Southern Power Grid Corporation
CSRC   China Securities Regulatory Commission
EIB   European Investment Bank
ETS   Emission Trading Scheme
FIT   Feed-in Tariff
FYP   Five Year Plan
gce   grams of coal equivalent
GDP   Gross Domestic Product
GDPD   GD Power Development Company
GHG   Greenhouse Gas
GW   GigaWatt
GWp   potential GigaWatt
HKD   Hong Kong Dollar
IPO   Initial Public Offering
IRR   Internal Rate of Return
kcal   kilocalories
kV   kiloVolt
kWh   kiloWatt hour
LNG   Liquefied Natural Gas
MBS   Mortgage Backed Security

MJ	Mega Joules
MW	MegaWatt
NDRC	National Development and Reform Commission
NEA	National Energy Administration
PV	Photovoltaic
R&D	Research & Development
RMB	Renminbi; Chinese currency
RO	Renewable Obligation
ROA	Return on Assets
RPS	Renewable Portfolio Standard
RQS	Renewable Quota System
SASAC	State-owned Assets Supervision and Administration Commission
SERC	State Electricity Regulatory Commission
SGCC	State Grid Corporation of China
SOE	State-owned Enterprise
SPC	State Planning Commission
SPCC	State Power Corporation of China
T&D	Transmission and Distribution
TGC	Tradable Green Certificate
TW	TeraWatt
TWh	TeraWatthour
UHV	Ultra-high Voltage
USD	US Dollar
W	Watt
WDI	World Bank World Development Indicators

LIST OF FIGURES

LIST OF TABLES

Part I

CHAPTER 1

China's Power Sector

Xinmin Hu

Abstract China's power sector has been the driving force for the country's economic and social development but also a cause of environmental pressures. Additionally, institutional features may undermine sustainability: institutional boundaries lack clarity, there is no level playing field for industry participants, information disclosure is often limited, and enforcement of regulations is not always rigorous. The sector's development has been uneven, with periods of rapid expansion against a backdrop of supply shortages and later growth slowdowns as overcapacity increases. While the legal framework behind the sector has expanded, institutional reform remains difficult given the tight relationships between party, state and industry. Tensions also abound within and between bureaucratic agencies. Despite its size and complexity, China has undergone many reforms and is poised for a low-carbon future.

Keywords Development • Generation mix • Industrial structure • Institutional structure • Policy

X. Hu (✉)
The Lantau Group, Kwai Fong, Hong Kong

© The Author(s) 2018
L. Lester, M. Thomas (eds.), *China's Electricity Sector*,
https://doi.org/10.1007/978-981-10-8192-7_1

AN INDUSTRY OF MIRACLES?

China's electricity industry has been the powerhouse for China's economic and social development. The electricity industry has been behind many of the country's modern miracles yet it has also faced a series of challenges relating to efficiency and development, and has been closely linked to China's environmental problems.

This book is an attempt to take stock of the world's largest electricity industry. As we look back over the sector's developmental history, assess the status quo and discuss its prospects for the future, we will take in both the miracles and the challenges. We will describe the constraints the sector must wrestle with. These include both hard constraints—such as resource location, abundance and consumption, environmental capacity and fragility, and industrial infrastructure—and soft constraints—such as the institutions, governance framework and cultures—that affect the industry's efficiency, how resources are used and allocated, and the available pathways to reform.

China is the world's second-largest country and has the world's largest population. In 2014, it became the world's largest economy by purchasing power; at 2016 exchange rates, China ranked second after the United States. The country's electricity sector has a scale to match. At the end of 2016, total installed capacity was 1646 GW and generation had reached 6142 TWh, both the largest in the world. As is so often the case, big means complicated, which in turn means that major problems can arise from seemingly small events, while incremental improvements in efficiency can lead to large overall energy savings. It is a power system that attracts attention and deserves proper examination.

But in contrast, China's endowment of traditional natural resources takes a more nuanced feel. True, the country has the world's third-largest coal reserves, but when looked at on a per capita basis, China has 116 per cent of the world's average for coal, 16 per cent for natural gas and just 8 per cent for oil (this despite being the world's second-largest oil consumer). What is more, these natural resources tend to be located far from the major centres of consumption.

Figure 1.1 shows the distribution of electricity demand and energy resources. Also shown is the Hu Huanyong Line, which divides China in two. The eastern side of the Hu Huanyong Line has less than half of China's land but nearly 94 per cent of the country's population and more than 94 per cent of China's Gross Domestic Product (GDP). Even within

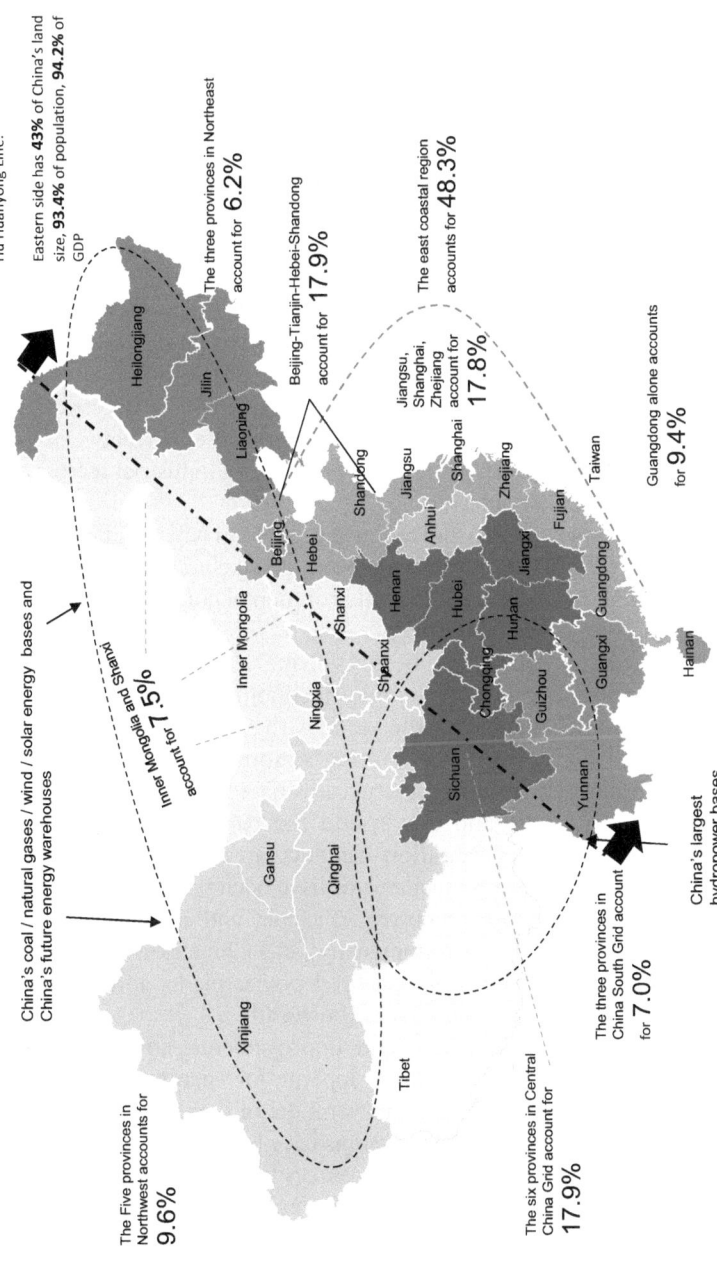

Fig. 1.1 A China of two parts—the locations of resources and electricity demand are far apart; percentages for each province or group of provinces show the share of total national power demand (Data Source: China Energy Yearbook 2016 (data for 2015))

Hu Huanyong Line:

Eastern side has **43%** of China's land size, **93.4%** of population, **94.2%** of GDP

The three provinces in Northeast account for 6.2%

Beijing-Tianjin-Hebei-Shandong account for 17.9%

The east coastal region accounts for 48.3%

Jiangsu, Shanghai, Zhejiang account for 17.8%

Guangdong alone accounts for 9.4%

Inner Mongolia and Shanxi account for 7.5%

China's coal / natural gases / wind / solar energy bases and China's future energy warehouses

China's largest hydropower bases

The three provinces in China South Grid account for 7.0%

The Five provinces in Northwest accounts for 9.6%

The six provinces in Central China Grid account for 17.9%

Heilongjiang
Jilin
Liaoning
Shandong
Jiangsu
Shanghai
Zhejiang
Taiwan
Beijing
Hebei
Anhui
Fujian
Shanxi
Henan
Hubei
Jiangxi
Inner Mongolia
Shaanxi
Chongqing
Hunan
Guangdong
Ningxia
Guizhou
Guangxi
Hainan
Gansu
Sichuan
Yunnan
Qinghai
Xinjiang
Tibet

this eastern area, population and GDP are much more concentrated in the six provinces and three municipalities along the coast. For example, this coastal region consumes more than 48 per cent of China's electricity. In contrast, coal and quality wind and solar resources are all largely located to the west of the Hu Huanyong Line, in the North, Northeast and Northwest (Three Norths); remaining untapped hydro resources are predominantly found in Yunnan, Sichuan and Tibet. The locations of these resources are typically at least a thousand kilometres away from the eastern coast. The implications of this are manifold and dramatic:

- The environmental capacity of the eastern coast is limited and is increasingly constraining further development.
- Efficient bulk transport of fuels and/or electricity is necessary and inevitable. This will require highly coordinated effort between various governments (both central and provincial), industrial sectors and power sector stakeholders.
- The development of transport infrastructure can have a critical impact on the interests and motivations of various stakeholders, giving rise to significant social, economic and environmental consequences.

China's electricity supply has been evolving alongside the country's developing economic structures, energy technologies and policies. The generation mix has changed over the years, with coal's dominance declining (though still remaining). The country's industrial structure has also been far from stagnant: ownership structures have changed, market concentrations have gone up and down, and the balance between vertical integration and market specialisation has fluctuated. In terms of regulation, pricing and tariff-setting mechanisms have been slowly aligning more closely with underlying cost structures; efficiency and emission standards for power generation have been improving, and China now operates the largest fleet of ultra-supercritical coal-fired power plants and ultra-high voltage (UHV) transmission systems in the world.

Despite these apparent changes and improvements, concerns remain over the sustainability of the country's appetite for energy, electricity and coal. Especially since 2003, the country's model of development has been resource intensive: environmental emissions have been high, and fossil fuels and other natural and environmental resources have been depleted. China produces and consumes half of the world's annual coal total and its use of hydropower has grown to the point where, in 2016, it was home to 30 per cent of the world's total hydro generation (up from just 3 per cent in 1980).

Concerns have grown over the wider sustainability of China's model of economic development as well. Critics have pointed to a number of potential bottlenecks or constraints that policymakers may now be trying to tackle in a more concerted manner than before.

- **Institutional boundaries lack clarity**. Mapping out exactly where government ends and companies begin is difficult not only because of state-owned enterprises but because of more widespread government and party involvement in both the country's economy and in the business operation of individual enterprises. In the past, the government has been active in picking winners and losers across the power sector and its supply chain, directly incentivising both companies and chosen sectors such as coal producers, coal-fired generators and renewable developers. The result has been a cycle of over- and undersupply, as inconsistent planning and improper responses to market conditions (through such things as distorted pricing mechanisms) have stymied the efficient utilisation of capital or resources and undermined reform. China is not alone in developing policies to aid the development of particular industrial sectors, and institutional reform need not mean that governments or regulators are to be less involved in managing or regulating the power industry. Instead, institutional reform is usually framed to focus on the establishment of clearly defined policy frameworks for rule-making and rule-enforcing, clarifying the roles of government and enterprise, facilitating a level playing field and helping to improve economic efficiency.
- **Not all companies face the same level playing field**. Regardless of the intent of policies, not all companies are treated equally, and this reduces fair competition and distorts market signals. Much of this inequality is rooted in the involvement of government in the operation of industry (through direct ownership of state-owned enterprises, preferential finance and explicit regulatory frameworks). The inequality may also be felt in other ways, such as in the market rules governing power exchange trading, which differentiate between technologies (coal/thermal versus others), or through the rights of state-owned enterprises, local-government-owned players and private players (again differentiated between domestic and foreign). For example, while thermal generation and renewable power technologies are different in terms of cost and environmental impact, these differences

should not serve as a basis to disqualify participation in markets or group them into different priorities; instead, externalities such as environmental impact should be internalised to a common competitive market accessible to all.

- **Information disclosure remains limited**. As market decisions are delegated to market players rather than central planners, it is crucial for the market to have access to relevant, accurate and timely data. Otherwise, market participants and potential market entrants will struggle to make informed decisions. In the absence of such information, it is hard for market participants and consumers to respond to changes in market conditions, be they to do with supply and demand or other system/commercial/policy events. Transparent information and operation also discourage rent-seeking behaviours, further improving economic efficiency. Transparency rules can also be extended to include sufficient consultation and grace periods for new policies and regulations.
- **Stable policy frameworks will be critical to China's low carbon transition**. Last but not the least, robust, foreseeable and stable energy policies and regulatory regimes are critical to an efficient transition to a lower carbon and greener energy industry and to maintaining future energy supply security and reliability, as the transition involves redistribution or reallocation of huge interests, benefits and costs. The total assets, in 2015, of the energy supply industry (coal, oil and gas, electricity and heat, processing of petroleum, coking and nuclear fuel) of Industrial Enterprises above Designated Size (those with an annual revenue from their core businesses of at least RMB 20 million) account for 23 per cent of the total assets from Industrial Enterprises above Designated Size. The enforcement of laws, regulations and rules must be strengthened, and the corresponding processes and procedures must be transparent.

As the complex engine behind China's economic miracle, just as much for its size and importance within global energy systems or for the challenges that it has yet to fully overcome, China's power sector is worthy of a closer look. This closer look is the subject of this book.

Intended primarily for university students and financial or business analysts seeking an introductory text to China's power sector, this volume has been designed to be accessible to all, whatever the reader's background, level of knowledge or specialisation. Despite China's obvious size and growing importance, too often too little is known of how it actually works. Ten years ago, many in the international energy industry were still belatedly waking up to China's coming of age; ten years on and for many, the opportunities and challenges in this largest of all markets remain obscure.

Helping companies and governments understand the intricacies of Asia Pacific's energy markets has been the core business function of The Lantau Group since it was founded in Hong Kong in 2010. From the start, we have followed developments in China, helping clients grapple with complexities around investment decisions as they have sought to understand the myriad policies and regulations, opportunities and pitfalls. For this book, The Lantau Group has teamed up with a group of academics with an unparalleled understanding of China's power sector. Whether well-established names in their field, or younger scholars still building their reputations, the academics who have contributed chapters to this volume all possess a very special connection to China's power sector, be it through scholarship, industrial experience, policy advice work or a combination of all.

The chapters that make up Part II of this book all focus on a particular aspect of China's power sector and are designed to be read singly or in sequence. Each presents itself as a stand-alone essay, properly referenced throughout, with a summary and highlighted key insights to help readers navigate the main arguments being presented. Yet each chapter also indirectly builds on its predecessors to explore a different aspect of the industry while contributing to an overall understanding.

Written by Philip Andrews-Speed from the National University of Singapore, the first chapter in Part II, Chap. 2, looks at the history and development of the power sector's system of governance. The nature of government in China's Party-State is explored, with explanations for what this has meant, and will continue to mean, for China's power sector. Readers will become aware of how political priorities have translated into changing policies and attitudes towards the electricity sector. They will be introduced to some of the most important regulatory and industrial stakeholders and introduced to the growing concerns around economic efficiency and environmental sustainability.

In Chap. 3, Xiying Liu, an independent energy economist, picks up where Chap. 2 left off with a more detailed look at the stakeholders of China's modern power sector. She also explores the current round of reforms that was kicked off in 2015 soon after the release of Xi Jinping's ambitions for reform were announced following the Third Plenum at the end of 2013. The current issues around electricity pricing are discussed, along with the nature of some of the compromises that have emerged out of the reform process. As the chapter makes clear, much has changed, and the power sector of today is in many ways very different from before, but the journey remains unfinished.

In Chaps. 4, 5 and 6 we move away from sector overviews and take more detailed looks at some of the topics that may prove most important for tomorrow: renewable energy, environmental legislation and finance. In each chapter, readers will be introduced to both the main developments made to date and some of the issues yet to be unwound. In Chap. 4, Sufang Zhang, from the North China Electric Power University, sets out the great achievements China has made in wind and solar generation (achievements that mean it now has the world's largest installed capacity of both), discussing both the country's *Renewable Energy Law* and the all-important reforming *Document No. 9*. But she also sets out the difficulties that the country still faces in curtailment of its low carbon energy and the physical, policy and institutional constraints that lie behind it. In Chap. 5, two authors from the Chinese Academy of Sciences take a more detailed look at the country's growing framework of environmental legislation, helping the reader to understand that the government has been highly active in trying to limit the power sector's environmental footprint. Here we look beyond carbon and renewables, and gain greater insight into how environmental policies across a whole spectrum of pollutants have helped shape the industry and fuel mix of both today and tomorrow. In our final guest chapter, researchers from China's Southwestern University of Finance and Economics show how, in the face of the enormous ongoing investment being directed into the power sector, the very nature of the country's finance and capital markets has changed, with renewable energy companies being among some of the most financially innovative in the industry.

Through these chapters, readers should be able to gain familiarity with the state of China's electricity market, its controversies and its emerging future trends. No short volume can answer all questions, but this volume should empower readers to know what questions to ask, and to understand the opportunities and risks that come with operating in China's developing electricity sector.

FROM SUPPLY SHORTAGE TO OVERCAPACITY

China's achievements in the power industry have been amazing in several areas. Installed capacity has grown from 1.9 GW in 1949, when the People's Republic of China was established, to 1766 GW by the first half of 2017, a growth of nearly 950 times in about 68 years. Power consumption grew from 623 TWh in 1990 to 5920 TWh in 2016. The average annual investment in the power sector was RMB 751 billion over the past ten years.

The expansion of the power industry has been pivotal for China's economic development. Not only has it helped fuel China as the world's manufacturing centre, earning China a huge trade surplus, but in turn, it has helped make China a global investing and buying power and enabled some of China's power related companies to grow enormously. The Big Five state-owned thermal power generation companies and two state-owned grid companies now stand in the Fortune 500 List (2017). State Grid is ranked second. Many of China's oil and gas, coal mining and engineering companies are also on the list.

There are four aspects of China that may be beneficial to understand:

1. Historical development path (cycles, restructuring institutions and planning, fuel mix, technologies);
2. Critical industrial issues (supply adequacy, pricing and dispatch, renewable energy, nuclear, the treatment of coal power, the role of gas, institutional reform, international collaboration);
3. Outlooks and visions for the future (uncertainties, challenges and opportunities); and
4. Specifics of ongoing power market reform.

Much of this is discussed in detail in Part II, but a brief introduction is presented here as part of a systematic introduction to China's power sector as a whole.

A simple (and common) view of China's power sector sees it following an uneven and complex development pathway, in which competing policies, institutional and industrial structures, regulatory and planning frameworks, and models of operation have competed with each over time to give rise to the modern sector. These political discussions have not ended. How to ensure the efficient and secure development of the power industry remains a critical challenge for the country's policymakers and industry stakeholders, just as it is for those in other countries. But within this paradigm lies a more complicated story.

China battled persistent power supply shortages before 1997. After opening its economy in 1978 to social and economic development, GDP growth rose by double digits from 1983 to 1988, with average annual growth of 11.88 per cent. In the five years from 1984 to 1988, the average was 12.09 per cent. In contrast, for various reasons including shortages of capital, technology and skilled labourers and engineers, the growth in installed capacity lagged behind: the average for the period was only 9 per cent; 5 per cent in 1984.

China first allowed foreign investors to build and operate power plants in China in 1985. The government wanted to bring in international experience and expertise in technology, design and operation in order to help tackle the expected high demand growth during a shortage of domestic capital. Average utilisation hours are shown in Fig. 1.2; following the decision, these trended down until about 1998, a clear demonstration that the right decision had been taken at the right time to help meet the growing demand. Foreign investors were not alone in being invited to invest in generation. Domestic investors, from central government to various levels of local government, were all encouraged to build power plants as well. In fact, some provinces, such as Shandong, Fujian and Guangdong, published a 'Three-Guarantees' policy (Minimum Utilisation, Minimum On-Grid Price and Minimum Return on Investment) to attract investment in power generation. Many other provinces and autonomous regions

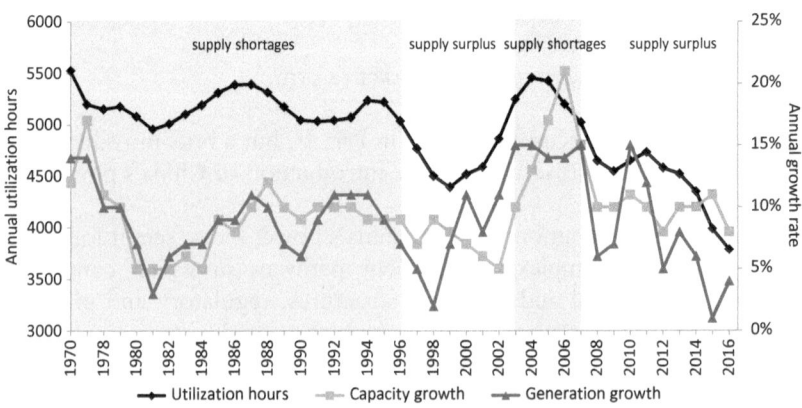

Fig. 1.2   Cycles of power shortages and surpluses

set up similar policies. These were very effective: by the end of 1998, in addition to domestic enterprises, there were 39 foreign companies invested in power generation. In 1997, they together held 12 per cent of overall capacity. Yet if the policies were successful in boosting generation, they failed to ensure consistent planning and approval systems or standards. As a result, many small and inefficient coal and oil-fired power plants were built without due regard for the availability of local versus imported fuel supplies, environmental consequences, or the logistics and costs of fuel purchase and transport.

When the Asian Financial Crisis hit in 1997, the power shortages had eased: capacity had caught up and both economic growth and demand for electricity had slumped. In 1997 and 1998, demand for power grew by only 4.4 per cent and 2.8 per cent, respectively: lows not seen again until 2015 (see Fig. 1.2). In response, the government halted any approvals of new thermal power projects for the three years from 1998 to 2001 and in 1999 started to retire the smaller coal-fired power plants. Given the lead time from planning and approval to construction and commissioning, this interruption in approvals led to capacity expansions slipping to 7 per cent in 2000, then to 6 per cent in 2001 and 5 per cent in 2002.

The Asian Financial Crisis did not act as a drag on China's economic growth for long. GDP growth steadily recovered to a level above 8.3 per cent after 2000 and stayed above 10 per cent from 2003 to 2007. Electricity demand also recovered and, given the slowdown in capacity additions, a new round of power shortages emerged after 2002. These shortages spread across China to multiple provinces including Yunnan, Sichuan and Hubei, following the rapid expansion of the country's infrastructure and housing stock, both highly reliant on inputs from energy (electricity) intensive sectors. In 2004, State Grid estimated a capacity shortfall of 30 GW.

The response was a new wave of capacity additions that gathered momentum after 2003 and has continued to the present day. In the period 2003 to 2016, only 2012 and 2016 saw growth in capacity fall below 10 per cent. The result was not only the banishment of power shortages by 2010 but also the rising spectre of overcapacity that has caused its own problems for generators.

In the background to this story, a number of additional changes affected the sector.

- From a peak of 14.3 per cent of total power investment in 1998, foreign investors began to quit China's power sector from 2003, with generation now overwhelmingly domestic.
- Generation technologies have become more diverse, more modern and more efficient. Many of the small and inefficient units of the 1980s and 1990s have gone, to be replaced with ultra-supercritical coal-fired generators, large circulating fluidised bed projects designed to improve the environmental performance of coal combustion, AP1000 nuclear units, large hydropower turbines, and bigger and better wind turbines. In addition, new and reinvigorated air pollution controls have been introduced as China operated a 'policy of a market for technologies'.
- Human capital has been improved across the entire sector. The country now enjoys a skilled labour force for equipment manufacturing, and plant design, engineering, commissioning, operation and maintenance.

Supply surpluses first started to emerge in 2009. The pain of this overcapacity has mostly been felt by coal-fired power generators nationwide, and wind and solar generators in the 'Three Norths' of China (the Northeast, Northwest and North), in the form of declining utilisation hours. For coal power plants, this decline has largely been driven by slowed demand growth, environmental policies imposing restrictions on coal consumption and competition from wind and solar as well as from hydropower and nuclear power plants. For wind and solar power plants in the 'Three Norths', the overcapacity is a product of relatively low levels of local demand coupled with operational arrangements limiting the ability of the renewable generators to export their power beyond the region.

The economic costs of both power shortages and surpluses are huge but also very hard to calculate accurately. The added value of electricity to an economy depends on many different factors, such as developmental stage and economic structure. Hunan province estimated the value to industry created per kWh of electricity at RMB 10 in 2004; the 2004 power shortages in the province resulted in direct economic losses of RMB 20 billion. On the other hand, the stranded value in surplus coal-fired capacity in China is estimated to reach RMB 2.45 trillion by 2020.

In an economy as large, complex and fast-growing as China's, it is near impossible to predict future demand accurately enough for centralized planning to ensure supply balances demand. Electricity demand depends on the speed of economic growth, changes in economic structure and technological development in both the power sector and industry at large. China's current supply surplus can be seen as a product of both the Asian Financial Crisis of 1997 and the Global Financial Crisis of 2007–09: major macroeconomic events that few accurately predicted.

Still, lessons can be drawn that highlight the inefficiencies in China's traditional system for planning and approvals that have been the root cause of recurring cycles of supply shortage and surplus.

- Good plans depend upon good information, but the quality of information in China is routinely questioned for both accuracy and timeliness. How strictly plans are implemented is also questioned, leading to a further source of uncertainty.
- The mechanism for allocating annual utilisation hours to power plants can distort market signals, disincentivising investors from preparing proper due diligence or risk assessments of investment opportunities. Since China does not have a transparent, competitive power market, market signals are often limited and indirect, and subject to manipulation.
- Investment decisions are not always taken for purely financial or commercial reasons. More strategic reasons, such as the availability of capital or resources, can also feature in decision making, as can market positioning and political concerns.
- As mentioned before, the opaque nature of the approval process has enabled rent-seeking behaviours to flourish.

The current state of ubiquitous overcapacity has major implications for the healthy operation of power enterprises, including the proper utilisation of capital, skills, and engineering resources and capabilities. Thermal power plants, in particular, are facing declining utilisation hours as slowing demand is coupled with increasingly stringent environmental controls.

INSTITUTIONAL REFORM: EASY TO SAY BUT HARD TO ACHIEVE

It has never been easy for China to set up appropriate governance structures among its national-level authorities, or between the central, provincial and local governments. At the national level, the Ministry of Power

Industry was established and abolished three times, and was merged with the Ministry of Water Resources three times, during the period from 1949 to 1998. The governance of the power industry was twice brought under a unified energy governance ministry, the first being the Ministry of Fuel Industry (1949–55), the second being the Ministry of Energy (1988–93), itself an amalgam of the then Ministries of Coal Industry, of Petroleum Industry, of Water Sources and Power Industry, and of the Nuclear Industry. In 2003, the State Electricity Regulatory Commission (SERC) was established to, among other responsibilities, plan the development and establishment of electricity markets, regulate and monitor their operation and propose legislation needed to establish regulations. However, SERC was abolished and incorporated into the National Energy Administration (NEA) in 2013 because reform of the electricity market had not progressed and there was no market for SERC to regulate.

China has clear procedures for enacting legislation, but the laws governing energy, and in particular electricity, fall far short of what is needed. Legislative progress started late and halted early. An *Electricity Law* was enacted in 1995, a *Coal Industry Law* in 1996 and an *Energy Conservation Law* in 1997, but it was not until 2006 that the *Renewable Energy Law* was implemented. A draft *Energy Law* completed its public consultation exercise in 2007 but there has been no subsequent progress and it remains off the statute books. There have also been a number of attempts to revise the *Electricity Law*, the current version of which is inadequate to support the need for institutional reform or the establishment of electricity markets.

These difficulties in governance have roots in China's ruling ideology, in the relationship between the Communist Party of China (CPC) and the government, and in the relationship between government and enterprise. These relationships have moulded the developmental pathway of the power industry's restructuring and reform. Control of various segments of the electricity industry has been subject to intermittent power struggles: control, or at least leadership, has granted the power to determine policy, institutional boundaries and systems of governance. The results have been industry stakeholders that combine rule-making with monitoring and enforcement while also being active participants. It has been a system, through its non-transparency, prone to unfairness and market distortion.

Just as with the ministry itself, a number of government agencies designed to manage and regulate the power sector have been set up and abolished. State-owned enterprises have been built up, dismantled and reconsolidated. For a country with China's reputation for centralisation and government control, the truth is surprising. China is a highly decentralised country, and this is clearly visible in the power sector. There is no single centralised regulatory agency for policymaking, despite various attempts to bring such powers together. Regulatory roles remain spread among a number of agencies, sometimes in conflicting manners.

While the National Development and Reform Commission (NDRC) is the major policy maker for China's macroeconomic policies, it also gets involved in technical issues of tariff setting. In addition, its semi-independent NEA acts as energy market regulator and project approver. The State-owned Assets Supervision and Administration Commission (SASAC), a ministerial-level government agency, is the owner of the central-level (non-financial) state-owned companies, including the two grid companies and all the national-level generation companies, major oil/gas companies and coal producers. SASAC, therefore, has key roles in both the strategic and operational levels of the power sector. It has incentives to influence national energy and power sector policy to the benefit of itself (as a ministry seeking an economically efficient industry), and on behalf of the companies under its administration (which seek higher profits and opportunities for expansion).

This tension within SASAC's roles reflects a fundamental tension within China's governance structures. As a country based on the nationalisation of private assets and the idea of public ownership, there are ongoing ideological debates on the merits of a 'socialist economy' versus a 'private economy'. Still a critical issue, the ebb and flow of power between the two ideologies can be followed in the history of the power industry itself.

- There are indications that the country's preference for central planning over free markets remains intact. During the on-and-off ten years (1955–58, 1979–82 and 1993–97) in which the Ministry of Power Industry existed, it both owned and operated power assets. Much of the argument that led to the repeated creation and abolition of ministries was not over the extent to which the government should exercise direct control of power assets, but rather over how the power industry should be fitted into the management of the broader fuel industry (including water resources, given China's use of hydropower). Unsurprisingly, the industry has not been well managed, and even now, there remain problems in how coal supply and price have affected power generation. This has been a particular issue in both 2016 and 2017. The preference for central planning is also evinced in the decision to merge China Shenhua Group with China Guodian Group.

- There are ongoing tensions between the management roles of government and industry. It is generally held that government should focus on policy and guide the direction of development, while industry should focus on operational issues. In the 1950s, five regional Bureau of Electric Power were set up to manage and supervise the power industry in Northeast China, Northwest China, North China, Central China and Eastern China. These partially succeeded in separating the two levels of management. In October 1988, the then Ministry of Energy published the *Reform Plan for Electric Power Industry Management Institutions*, which laid out plans to restructure regional and provincial electric power bureaus into corporations. This trend to separate out management roles has continued through further corporatisation.

- The State Council's 2002 *Reform Plan of Electric Power Industry Institutions*, commonly called *Document No. 5*, laid out further plans for reform of the power sector. There was to be a separation of generation from the grid, dispatch was to be based on offered prices and competition was to be introduced through the breakup of monopolies. The plan was only partially implemented. The State Power Corporation of China (SPC) was broken up into five generation companies, two grid companies and several other firms covering engineering and construction. Little else was achieved: dispatch was still centrally controlled, and competition was highly limited. In March 2015, the CPC and State Council released a new reform plan: *Further Strengthening the Institutional Reform of the Electric Power Industry*, also known as *Document No. 9*.

The tensions and compromises in China's approach to governance often reveal themselves in inefficient or inconsistent operational rules. A classic example lies in the treatment of coal and on-grid electricity prices. From 1993 to 2013, China used a dual-track pricing system for coal, with guiding prices and price caps for key coal users (including the major thermal generators) alongside market-determined prices. Typically, such a mechanism protected state-owned thermal generators from market coal price risks. In December 2004, China started using benchmark on-grid tariffs for coal-fired generators, and designed a power-coal price linkage mechanism to enable coal-fired on-grid electricity prices to adjust to changes in coal prices according to a pre-defined formula. Three versions of this linkage mechanism were released in 2004, 2012 and 2015. The fundamentals remained the same: the mechanism was an attempt to manage the winners and losers in changing market environments. However, the mechanisms have not been fully implemented: coal prices have exhibited ups and downs (generally rising from 2004 to 2011, then falling from late 2013 to early 2016, then growing in late 2016 and fluctuating at high levels due to NDRC policies restricting coal production) that have not been fully passed through to power prices.

When first implemented in May 2005, the benchmark pricing mechanism led to an average increase of RMB 0.018 per kWh for coal-fired on-grid tariffs. The first six-month update window passed without any change in the benchmark, but in May 2006, a further uplift in coal-fired on-grid tariffs of RMB 0.0088 per kWh was allowed. Two further price increases were allowed in 2008, together leading to an average increase of about RMB 0.037 per kWh. Despite a few later price adjustments, the benchmark price linkage mechanism was not used again given persistent concerns that the price rises were fuelling inflation. Instead, in 2008, the major thermal generators were allowed to expand their businesses into coal mining, creating vertically integrated enterprises in an attempt to dampen coal price risks. Then, in 2012 and the years following, SASAC and the Ministry of Finance provided direct cash injections to each of the Big Five thermal generators to help them deal with deteriorating financial conditions resulting from rising coal prices and associated climbing levels of debt.

Such actions illustrate the continuing nature of market distortions created by the current system of governance. In an attempt to maintain the viability of thermal generation while preventing pass-through of costs to end-users, the Big Five state-owned generators have received preferential treatment. This is just one of the ways in which the current governance and institutional framework skews competition and undermines proper market discipline surrounding investment and operation. It also reveals how China often acts contrary to Western expectations even in its direct interventions. Rather than separating 'bad' assets from the 'good' in an attempt to prevent contagion, China has often sought to protect the viability of 'bad' assets by injecting them with 'good' assets. This was seen clearly in the recent merger of Shenhua Group (a coal producer) with China Guodian Group (one of the Big Five thermal generators). The resulting vertically integrated giant has affected the competitive landscape for companies not fortunate enough to receive such state backing.

AN EVOLVING GENERATION STRUCTURE

China is endowed with reasonably abundant coal but much more limited natural gas and oil. There is significant hydropower potential, especially in Southwest China, Hubei and Hunan in Central, Qinghai in Northwest, and Jilin and Heilongjiang in Northeast China. China also has large wind and solar potentials, though the best quality resources are located in the Northwest, Northeast and North, which are far away from the demand centres in eastern and central provinces.

Figure 1.3 shows how national fuel mixes tend to be determined by what resources are available (either domestically or easily imported), what resources are affordable and how well developed renewable energy policies are. Generally, Asia Pacific uses more coal and less natural gas than other regions, while the Middle East takes the other extreme and is almost entirely dominated by oil and gas. China's fuel mix is much more heavily dependent upon coal (and much less dependent on oil and gas) than most of the rest of the world. This remains the case despite significant expansion of cleaner and lower carbon energy sources such as hydro, nuclear and renewables over the last decade.

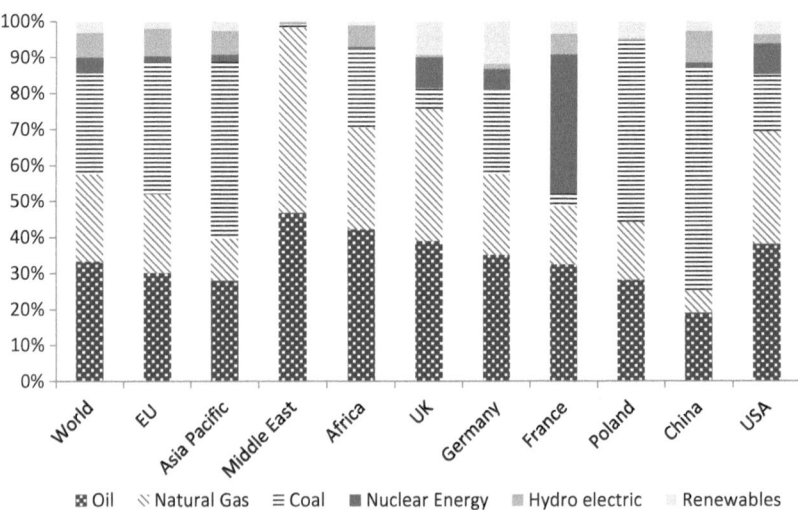

Fig. 1.3 Fuel mix of primary energy consumption in 2016 for selected regions and countries (Data source: BP Plc, Statistical Review of World Energy, June 2017)

The growth in low carbon alternatives to coal has created an additional source of pressure for traditional thermal generators. As power demand growth slows and pro-renewable policies bite, coal has begun to lose out. However, coal is unlikely to disappear in the medium term. Just half of the coal consumed in China goes to thermal generation where the environmental standards are most stringent. The other half is used in direct combustion, often in small and inefficient boilers. Continued electrification of China's economy could cut pollution while preserving levels of coal use. Even without this, coal's cost advantage will make its complete demise a slow one.

Before 1994, China's generation fuel mix was almost entirely thermal (coal, gas and oil) and hydro. In the years 1975 to 1993, hydropower was a very important source of electricity, responsible for between 24 and 32 per cent of installed capacity before entering a steady downward trend as thermal (especially coal) generation expanded massively. Until more recently, the share of generation from thermal capacity hung at around 80 per cent, though two nuclear reactors (in Zhejiang and Guangdong) were started in 1994 and wind power began to be built after 2000 (picking up speed in 2006). Solar power generation is a more recent addition still.

By the end of 2016, the picture had changed again.

- The installed capacity of gas, wind and solar had increased to 4.3 per cent, 9.0 per cent and 4.7 per cent, though corresponding generation shares were much lower, at 3.2 per cent, 4.2 per cent and just 1.1 per cent.
- The share of coal-fired generation had fallen to 67.4 per cent; absolute generation from coal has plateaued since 2013 at 4000 TWh a year, thanks to the addition of non-coal generation combined with the various environmental policies and direct controls on coal consumption.

Just as for other countries, meeting electricity demand growth through the least cost technology and fuel has been a guiding principle for China. This has been achieved by adapting or developing domestic generation and transmission technologies and using local resources. In the process, environmental protection and low carbon technologies have become increasingly important.

The speed with which China has had to expand its power supply is phenomenal: the average annual growth rate from 2000 to 2014 was above 10 per cent, and electricity consumption grew from 1347 TWh in 2000 to 5920 TWh in 2016. To meet this, annual capacity additions averaged 88 GW a year from 2005 to 2012, equivalent to nearly three 600 MW power plants a week. Much of this was coal-fired. Capacity expanded even more rapidly in the years from 2013 to 2016, with average annual additions of 124 GW.

To help meet the growing demand for power, the central government, in 1999 and 2000, proposed the 'West to East Power Transmission' programme. The plan was to establish major power generation bases in the west and then transmit the power eastwards towards the coastal and central demand centres. This was to have the benefits of boosting the GDP of the poorer, less-developed western provinces, while also relieving environmental pressure on the more highly developed (and higher-cost) eastern provinces. Three transmission corridors were to be established: the first from Shanxi, Inner Mongolia, Shaanxi and Qinghai to Beijing, Tianjin and North Hebei; the second from Hubei, Sichuan and Anhui to the Yangtze River Delta; and the third from Guizhou, Guangxi and Yunnan to Guangdong. Large hydropower bases were planned in the upper Yellow River, Yangtze River and Pearl River, along with many of Yunnan's rivers. Coal power bases were planned for Inner Mongolia, Shanxi, Shaanxi and Ningxia.

Lying behind this programme are substantial improvements in China's generation and transmission technologies, in terms of diversity of technologies, generation structure, plant unit size and power system efficiency.

- China commissioned its first ultra-supercritical generation unit in 2006, a 1000 MW unit at Yuhuan Power Station in Zhejiang. At a time when the national average was 370 grams of standard coal equivalent per kWh, Yuhuan consumed only 285.6 grams per kWh. By the end of 2016, there were ninety-six 1000 MW ultra-supercritical units in operation across China.
- To help transmit electricity efficiently from power bases to demand centres, often involving distances of over 1500 km, an extensive network of UHV transmission lines was constructed. By March 2017, 14 projects had been successfully commissioned, with a combined length of almost 19,000 km and capacity of 94 GW; a further 7 projects yet to be completed will extend this to 30,000 km.
- Two programmes designed to improve energy efficiency and reduce the power sector's environmental footprint were the 1999 'Shutting Down Small Thermal Power Plants' and 2006 'Replacing Small Units with Larger Ones' programmes, the first targeted power plants 50 MW and smaller. Between 1999 and 2015, 118.3 GW of small

and inefficient capacity was permanently retired, most of which was coal-fired (oil-fired units were also targeted, though these accounted for less than 8 per cent). A further 20 GW of inefficient coal-fired capacity is planned to be shut down in 2016–20 according to the *13th FYP Development Plan for the Electric Power Industry*.

- At the end of 2016, units of 1000 MW or above accounted for 9.22 per cent of China's thermal generation capacity, units of 600+ MW accounted for 32.57 per cent and those of 300+ MW accounted for 37.22 per cent. The shift towards larger and more efficient generators has seen coal use per kWh fall from 471 grams (standard coal equivalent) in 1978 to 392 grams in 2000 and 312 grams in 2016. The changed thermal generation structure has also greatly improved the environmental performance of the electricity sector. For example, thermal generation flue gas emissions of PM (particulate matter: smoke and dust), SO_2 and NO_x have been reduced from 1.6 grams, 5.57 grams and 3.04 grams per kWh in 2006 to 0.08 grams, 0.39 grams and 0.36 grams per kWh in 2016. Carbon intensity in 2016 had fallen to 0.822 kg per kWh, a drop of 21.6 per cent compared to that of 2005.

- The *Action Plan for Energy Conservation, Emissions Reduction, Upgrading and Retrofitting of Coal-fired Power Plants* (2014–16) set out plans for further reductions in heat rates (coal use per gram) and emissions caps, detailing both enforcement and timelines. A number of incentive policies have also been implemented to support the development of wind, solar and biomass. These include Feed-in Tariffs and a potential green electricity certificate trading scheme. A major challenge for China has been to set a logically consistent framework and detailed implementation plan to allow generators and end-users to internalise the costs of environmental externalities. Without such systematic and detailed plans and models, policymakers are vulnerable to unintended inconsistencies or contradictions between policies on renewable targets, Feed-in Tariffs and environmental levies.

A Growing Global Player

From relying on international experience and expertise, China has become a global player with its own research and development output, its own technological enhancements and its own experts. Many Chinese companies, especially the state-owned energy enterprises, have accumulated great experience, knowledge and capital. With significant overcapacity at home, the country is now looking to export its industrial expertise across the world, and many energy companies have been actively exploring overseas markets, both developing and mature.

China's first 'Going Abroad' policy was in 2000, after the Asian Financial crisis but before it joined the WTO in December 2001. Despite various incentivising policies, power companies were not much involved given their lack of experience in operation, construction or design. Since then, the power sector's international involvement has been mainly limited to equipment export or hydropower projects, often supported through bilateral government arrangements or international aid agencies. However, participation has now spread along the whole supply chain under a variety of business models including EPC, BOT, BOOT, PPP and PMC. Exported generation technologies have diversified from hydropower to gas-fired, coal-fired, oil-fired, nuclear, wind, solar, biomass and tidal generators. As China's domestic power sector technologies have improved, so the country has developed a growing reputation in international markets, especially with its ultra-supercritical generators and UHV transmission. Recent overseas contracts include the 2015 contract to build Indonesia's Java No.7 Power Project (2 x 1000 MW ultra-supercritical units) and a 2014 contract for a UHV transmission project in Brazil. Target markets have also expanded to include advanced economies such as the United States, Canada, Australia, Singapore, the United Kingdom and European Union member states.

International involvement has helped Chinese companies better understand what their competitive advantages are. As companies seek to position themselves as truly global competitors in international energy markets, they may increasingly look beyond opportunities established through government influence such as those linked to President Xi Jinping's flagship Belt and Road Initiative or the proposed global electricity network proposed by Liu Zhenya, chairman of State Grid. Unfortunately, international involvement has also led to trade disputes (such as those surrounding solar panel exports to the EU), local labour disputes and continued environmental concerns. There has also been international unease over some of the financing behind certain projects. Still, increasing overseas participation by Chinese enterprises may well lead to improvements in the transparency of both domestic operations and the relationships between commercial enterprises and any non-commercial owners.

A LEGACY OF REFORM

China's initial power industry restructure followed the international wave of power industry deregulation in the 1990s by only a few years. SPC was established in January 1997 in an attempt to separate business operation from government ownership and regulation. SPC initially operated the assets under the old Ministry of Power Industry; when the Ministry of Power Industry was abolished in 1998, regulatory powers were transferred to the State Economic and Trade Commission.

In 1998, six provinces ran pilot power market reforms that separated generation from transmission and distribution and established a competitive generation pool (though only open to approved power plants). These reforms were largely managed by SPC, as policymakers tried to take advantage of a supply surplus to rebalance an array of generator-specific on-grid tariffs, many of which had been approved and set during periods of power shortages. An additional strong driver was the need to further institutional reforms separating government and business.

The direction of reform changed following an investigative article published in April 2000 on the financial losses of Ertan Hydropower Station (3300 MW) in Sichuan Province. The plant had started construction in 1991 when Sichuan (which then included Chongqing) had strong demand growth and serious power shortages. Ertan Hydropower Station was intended to supply the Sichuan and Chongqing markets, but by the time of commissioning, both were in surplus and the station no longer had a viable market. It was argued that the real problem facing Ertan Hydropower Station was not local overcapacity, but policies that prioritised supply balancing within individual provinces and prevented power being sent out into larger regions. In October 2000, the Electricity Institutional Reform Coordination Group (EIRC) took over leadership of power market reforms. EIRC was led by the State Economic and Trade Commission, the Ministry of Finance, the Law and Legislation Office of the State Council, the Institutional Reform Office of State Council, SPC and China Electricity Council (CEC); the change was at least in part a result of an institutional power struggle.

In May 2002, the State Council released the *Reform Plan of Electric Power Industry Institutions*, setting the direction of China's power market reform for the next few years. Separating generation and auxiliary businesses (design, engineering, construction and equipment manufacturing) from the grid, and creating competitive markets and market-based pricing mechanisms were back on the agenda. Little happened though, and after 2005, reform stalled as power shortages once again began spreading across the country. It wasn't until after the supply situation eased in 2013–14 that reform restarted.

In March 2015, CPC and State Council released a document entitled *Further Strengthening the Institutional Reform of the Electric Power Industry*. By November 2015, six supporting documents had been released by the NDRC and NEA. *Opinions on Advancing the Creation of Electricity Markets* elaborated on the key components of an electricity market system; other documents detailed other subsectors and specific areas, such as transmission and distribution pricing reform, the orderly opening up of electricity generation markets, the opening up of the retail business and the administration of captive generation power plants. In August 2017, the NDRC and NEA released a *Notice on carrying out electricity spot market pilot projects*, which set out plans to implement spot markets in eight selected provinces/regions.

These policy documents marked an important step forward in laying out a road map for electricity market reform along with details of implementation. Unfortunately, the documents lacked an in-depth and forward-looking examination of the policies and, as a result, the current plans include a number of potential conflicts and inconsistencies.

- Despite a clear aim to increase market forces within the power sector, there are few details on how to reform the industry structure to enable a level playing field for market participants. Abuse of market power can undermine the successful implementation of markets, but the potential always exists when the market has dominant legacy players. A common way of dealing with such market concentration is to unbundle vertical services and divide big players up into smaller ones. The recent merger of China Shenhua Group with China Guodian Group is the opposite of this, building up a new vertically integrated player and entrenching market power.
- While the retail business has been opened up to new players, participants without their own distribution assets do not have metering or invoicing rights with their customers. This once again hands market power to legacy transmission and distribution companies.
- Provincial governments play a critical role in the design, establishment and implementation of their electricity market, but given current overcapacity, local governments have incentives to protect their

own generators from outside competition or to force generators to offer lower prices in favour of local industry. Such local vested interests can stymie the development of effective inter-provincial power trading, in turn undermining the intent behind the creation of power exchange centres in Beijing and Guangzhou.

- Dispatch merit orders have been a frequent topic of policy initiatives and were once again stressed in the August 2017 release of the *Notice on carrying out electricity spot market pilot projects.* Wind and solar power have been given first priority in dispatch in an attempt to boost renewables, but this conflicts with the common dispatch principle of spot markets where the merit order is based on offered prices and subject to system constraints. This may lead to potential inconsistencies in market design and implementation.

The story of reform in China has not been one of stasis, but nor has it been of constant progress. Rather, reforms have started and stopped, *started and stopped.* In an industry long held to be central to the success of China's economic success, it is hardly surprising that politics and competing ideologies and theories should intrude. As different leaders, vested interests and groups have come and gone, so have different aspects of reform prospered and stalled. Most often, reforms have moved forward most quickly when conditions have made it easy, such as when there is a supply surplus, and stalled most quickly when conditions have been more challenging, such as with supply shortages or rising prices.

At the heart of the tensions in power sector reform are the tensions in China's broader system of governance, and in this, the contrast with Western countries is shown most starkly. Yet China has accomplished much, and despite the many challenges ahead, the story is far from over.

Part II

CHAPTER 2

Governance of the Electricity Sector

Philip Andrews-Speed

Abstract This chapter shows how key elements of the governance of China's electrical power sector have evolved over the past twenty years. It will focus on such issues as policy paradigms and priorities, government and industry structures, the distribution of power and influence, the balance between administrative and economic instruments, and the links with trends in the wider economy and other sectors. The key political and economic importance of the power sector has meant that the government remains deeply involved in many aspects of the sector, from enterprise ownership to tariff setting, and has allowed market forces to play only a limited role. The recent round of reforms launched since 2015 marks a further incremental step in enhancing the role of market forces.

Keywords China • Electricity • Governance • Reform • Policy

P. Andrews-Speed (✉)
Energy Studies Institute, National University of Singapore,
Singapore, Singapore

SUMMARY

China has the world's largest electricity sector, in terms of capacity and output, and generating capacity now massively exceeds what is required to meet demand. This success has been achieved through the direct involvement of the state, in direct contrast to the Washington Consensus dogma of privatisation and liberalisation. The state retains majority or total ownership over most companies in the electricity supply chain, is deeply involved in planning and investment approvals, and continues to set or guide producer and consumer tariffs. This preference for state control arises from the critical importance that the Chinese Communist Party, as the government, attaches to the power sector in order to support economic development and secure the authority of the Party. Until 2004, the long-standing policy paradigm for the sector also embraced the priority assigned to investing in generation capacity rather than energy efficiency, and in taking advantage of the nation's ample supply of cheap coal.

This approach to governing the electricity sector has had two major consequences. First, reform has proceeded more slowly than in other sectors. Transmission and distribution infrastructure over most of the country is owned by one vast company, the State Grid Corporation. Despite the corporatisation of generating companies, economic competition on the basis of cost does not exist; rather, dispatch is planned and highly politicised. Finally, tariffs bear little relation to the balance between supply and demand. The reforms announced in 2015 are intended to change this situation. The second consequence of this deep state involvement takes the form of economic and environmental cost. The economic cost arises from the inefficiencies in the investment decision-making and management processes of the companies. The environmental cost takes the form of continuing air pollution from the burning of coal and the high level of carbon dioxide emissions, though the contribution of the power sector to this pollution has declined in recent years.

The first decade of this century saw two significant changes to this approach. The energy supply crisis that started in 2003 forced the government to instigate a vigorous programme to reduce national energy intensity. This involved closing old, small and inefficient thermal power plants, and constructing some of the most modern plants in the world. This shift of policy was soon supplemented by growing concerns over global climate change and nationwide air pollution. The combination of the continued search for security of energy supply and the new desire for clean energy led to a surge in the installation of non-fossil fuel electricity generating capacity, notably wind, solar PV and nuclear.

The reforms announced in 2015 mark a further incremental step in enhancing the role of market forces that will be accompanied by the introduction of a nationwide market for carbon emissions. Whilst these initiatives will probably yield some economic and environmental benefits, the gains are likely to be marginal for as long as state-owned enterprises (SOEs) continue to enjoy soft budgetary constraints and access to generous finance, and local governments retain influence over investment and dispatch decisions.

Key Insights

- The opportunities for foreign companies to invest in China's electricity infrastructure, such as generation or transmission and distribution, remain unattractive on account of the power of the incumbents and long-term policy ambiguity.
- The ongoing market reforms could provide opportunities for foreign companies to provide consulting and technical services to domestic market participants.
- The progress of reforms will depend on the higher-level political and economic priorities of the leadership. Which is more important: short-run economic efficiency, state control of a strategic industry, cheap electricity supply or clean energy? How this plays out will depend not just on economic ideology, but on competing political and economic interests.

Introduction

China's electricity sector is the largest in the world in many respects: total installed capacity, thermal and renewable energy capacity, and annual generation and consumption. It also has some of the most modern coal-fired plants in the world, and the fastest growing programme of nuclear power. At the same time, the nation's companies have become world leaders in the manufacture and export of renewable energy equipment, are heavily involved in the construction of hydroelectric dams around the world and aspire to export nuclear energy technology. What we see today is a far cry from the early decades of Communist Party rule when endemic power shortages undermined economic growth and standards of living.

Whilst much has changed, the power industry in China has remained more firmly under government control than other energy industries such as oil, gas and coal. The state, at central or local levels, is a majority or sole owner of most electrical power enterprises, most notably the giant State Grid Company, and has continued to set producer and consumer tariffs. Despite the corporatisation, unbundling and commercialisation of the industry, no true economic competition exists for either generation or retail. This is in the process of changing as a result of plans announced in 2015.

The aim of this chapter is to show how key elements of the governance of China's electrical power sector have evolved and, at the end, to provide an assessment of likely directions of future change. It will focus on such issues as policy paradigms and priorities, government and industry structures, the distribution of power and influence, the balance between administrative and economic instruments, and the links with trends in the wider economy and other sectors. The chapter begins with a survey of the wider political and economic context of the electricity sector, before moving on to examine, in turn, the general aspects of sector governance, structural reforms and policy instruments. This account ends with the announcement of further sector reforms in 2015, which is the topic of Chap. 3 of this book.

Political and Economic Context

The death of Chairman Mao in 1976 and the rise to power of Deng Xiaoping allowed China to move into a new era of politics and economic policy. The mode of political leadership changed from one based on individual personality to being more collective, at least among elites, and economic policy-making became more pluralised. Nevertheless, the Communist Party has remained at the core of the Party-State, and sustaining Party legitimacy and power continues to be the overriding imperative for the leadership.

Whilst private ownership and free markets were not yet terms used in formal government rhetoric, the 1980s saw a remarkable transformation of the economy in a number of ways: land was decollectivised; enterprises outside the formal state-owned sector were allowed to blossom, notably the township and village enterprises that were key drivers of economic growth at this time; and prices for some goods began to more closely reflect market forces. Though liberalisation was not a priority of the leadership at that time, modernisation was. This involved three key reforms to the governance of the economy: progressive delegation of authority to the provincial and lower levels of government; diversification of ownership of small- and medium-sized enterprises, including outright privatisation; and corporatisation of large SOEs. These changes led to the steady transformation of the system for economic policy-making and implementation from one that was, in principle if not in practice, highly centralised and planned, to one which became characterised as 'fragmented authoritarianism' [1]. The system became more fragmented as power was delegated downwards and as the number of government and corporate actors increased. Yet, its authoritarian character persisted through the sustained influence of the Communist Party.

The 1990s witnessed the steady increase in the role of market forces across many sectors of the economy. However, the energy sector remained an exception to these trends of privatisation and market liberalisation, as did other 'pillar industries' that were seen as being key to the nation's economic development and thus to the authority of the Party. Although the government restructured the energy sector and encouraged the corporatised enterprises to list minority stakes on domestic and international stock exchanges and to internationalise their businesses, the state at central or local levels retained controlling stakes. In a similar manner, energy prices for producers and consumers have been liberalised more slowly than in other sectors, and even today, energy prices for households remain either controlled or 'guided' by the government.

The importance of the energy sector arises from not just the need for energy to support economic growth, but also from the role of the SOEs in this sector as major sources of employment and, at times of high profits, of tax revenues. Only recently have SOEs managed to unload many of the wider social responsibilities onto the government. As a consequence, the central government still retains significant control over policy, industry strategies and major investments in the energy sector. From these observations, it is clear that the policy paradigm for the electricity sector requires state control in order to secure economic development and sustain Party authority.

The relatively weak role of market forces and the caution shown by the central government in raising energy prices, especially to households, mean that most policy instruments in the energy sector continue to be administrative in nature and take the form of targets and other obligations supported by a mix of financial support from state-owned banks and government agencies. Examples in recent years have included targets for energy intensity reduction imposed on enterprises and local governments, national targets for the installation of renewable energy, energy efficiency standards for appliances, product standards for oil refineries, emission standards for thermal power plants and obligations to close outdated industrial plants. Economic penalties for non-compliance with these administrative policies have tended to be weak, as have incentives for compliance. This has arisen from a reluctance to force companies into bankruptcy and increase unemployment. In recent years the government has been making greater efforts to constrain energy use and pollution. It has not only implemented administrative approaches more forcefully, but has also been increasing the deployment of policy instruments, such as prices and taxes to support energy or related environmental policy priorities. Most notably, the campaign to reduce energy intensity that ran from 2005 to 2010 saw a substantial increase in electricity tariffs for the industrial and commercial sectors. In addition, the government has launched a number of pilot carbon markets.

The number of actors involved in China's energy supply chain is large. In the days of economic planning, the State Planning Commission (SPC) was responsible for planning energy supply and demand, as well as investment, and for setting energy prices. Today, the National Development and Reform Commission (NDRC), in part through its National Energy Administration (NEA), continues to produce five-year energy plans, though it no longer controls supply and demand. The five-year plans for energy build on the national five-year plans and set out the overall objectives and targets for the sector. The NEA retains oversight of investment in the energy sector, though authority has been delegated recently to the provinces, continues to set some energy prices and retains the role of formulating and overseeing the implementation of key energy policy initiatives. Several other agencies are also involved in the governance of the energy sector, the most important of which are (see Fig. 2.1):

- The State-owned Assets Supervision and Administration Commission, which represents the government in its role as owner of the SOEs, overseeing the performance of the SOEs and their senior management;
- The Ministry of Environmental Protection, which sets and implements environmental regulations and standards;
- The Ministry of Land and Resources, which governs the management and use of land and natural resources including oil, natural gas and coal; and
- The Ministry of Science and Technology, which supports science and technology research and development.

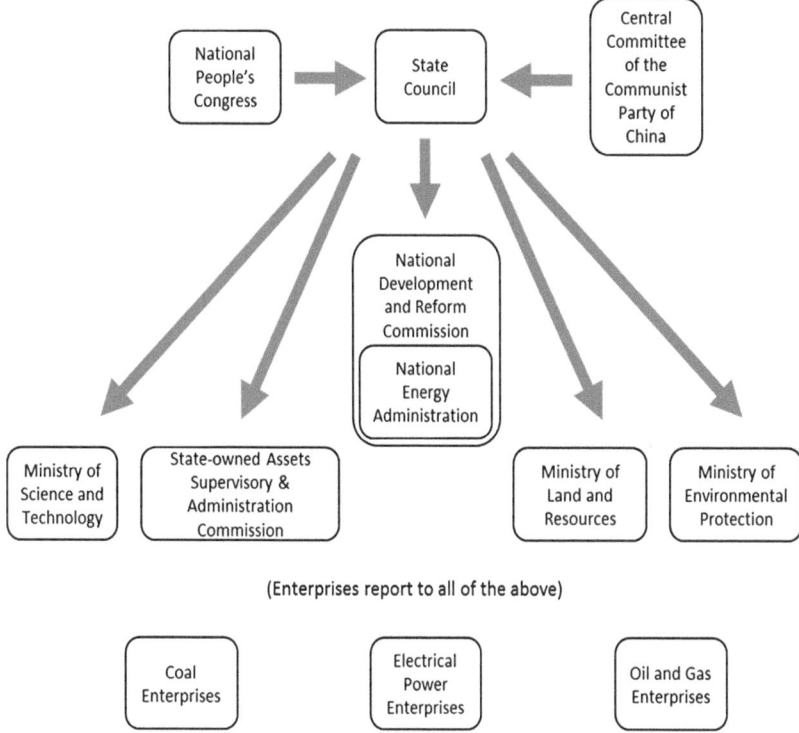

Fig. 2.1 Simplified scheme showing the main energy-related organisations and enterprises at central government level in 2017. All organisations have local bureaus or subsidiaries at provincial, prefecture and county levels

Other relevant agencies include:

- The Ministry of Water Resources, which governs the use of hydroelectric resources;
- The Ministry of Finance, which is responsible for fiscal policy and economic regulation, among other matters;
- The Ministry of Industry and Information Technology, which sets standards and issues regulations for industry; and
- The Ministry of Commerce, which oversees foreign trade and investment.

A National Energy Commission was created in 2010 but it is deemed to have been largely ineffective. Leading Groups within the State Council that are chaired by the Prime Minister of the day provide some level of coordination. Despite these moves and the growing role of the Communist Party's Leading Small Groups, the governance of energy remains fragmented [2]. At the same time, there is no clear separation of policy-making, planning and regulation [3].

The NDRC, the NEA and the other various agencies all have equivalent bureaus at provincial and lower levels of government that are charged with adapting and implementing central government policies. However, the continued focus of local governments, particularly at the county level, on economic growth has led them to dilute, distort and even undermine central government directives, such as those relating to environmental protection, workplace safety, energy efficiency and clean energy [4]. Only recently have steps been taken to reduce the emphasis on Gross Domestic Product for the career incentives of local government officials. Conversely, if the central government policy is aligned with the interests of local governments, then the policy can be implemented with such vigour that it produces undesirable, unintended consequences. An example is the reported temporary closure of factories and other organisations by some local governments in late 2010 in order to meet energy intensity targets.

The other key policy actors are the SOEs in the coal, electricity, and oil and gas industries. Despite partial privatisation through initial public offerings, these enterprises retain close links with government. The closeness of this relationship arises from the origins of these enterprises as government ministries, the personal connections between individuals that result from the promotion of enterprise managers into government and the ongoing importance of the energy sector to China's economy [5]. Overarching these factors is the powerful role that the Communist Party plays in overseeing the appointment of senior individuals in both government and enterprise through the *nomenklatura* system. Whilst this closeness can enhance coordination in policy processes, it also allows both sides to take advantage of the other. The government continues to impose non-commercial objectives on energy SOEs, not least through pricing mechanisms that result in financial losses and by discouraging them from laying off workers. Conversely, the enterprises are able to exert influence over policy-making through their control over information and the fact that their Chief Executives have ministerial rank. These phenomena are replicated at lower levels in the hierarchy, for many energy enterprises are owned by local governments.

The end-users of energy include industrial and commercial enterprises, the agricultural sector, as well as households and individuals. The historic importance of energy-intensive heavy industries, such as steel, non-ferrous metals, chemicals and cement, has meant that the government has always taken into account their interests, most notably by keeping their energy tariffs at relatively low levels. This has changed since 2004 because of efforts to reduce national energy intensity and pollution that included raising electricity prices and closing outdated industrial capacity. Nevertheless, the central government continues to address the interests of the agricultural sector and households by keeping electricity tariffs low, despite the introduction of tiered tariffs for households.

GENERAL FEATURES OF SECTOR GOVERNANCE

The need for an adequate and reliable supply of energy, especially of electricity, has been a top priority for China's government since the Communist Party took power in 1949. Two priorities brought energy onto Mao's agenda: the need to support rapid reconstruction and industrialisation [6] and the desire to electrify the countryside [7]. When the Peoples' Republic of China was founded in 1949, the total electricity generating capacity in China amounted to less than 2 GW and the capacity of small-scale hydroelectric plants to support the nation's massive rural population was just 33 MW [7, 8]. At that time the population was about 540 million. In comparison, the generating capacity in the USA was 63 GW in 1949 for a population of 150 million [9].

The death of Mao in 1976 and the subsequent rise to power of Deng Xiaoping triggered a new round of industrialisation that needed a sustained increase in electricity supply. The government delegated more power for planning and financing electricity infrastructure to provincial governments and introduced a two-tier tariff mechanism for power plants: output that was planned would be sold at regulated prices, but output that exceeded the plan could be sold at market prices. This allowed end-user tariffs to rise gradually [10]. These and later reforms supported a programme of investment which drove total installed generation capacity from 45 GW in 1980 to 167 GW by 1992 and then to 357 GW by 2002 (see Fig. 2.2).

The push for self-sufficiency and the abundance of the resource endowment have supported the long-standing dominance of coal in the fuel mix for electrical power. The abundance of cheap domestic coal resources

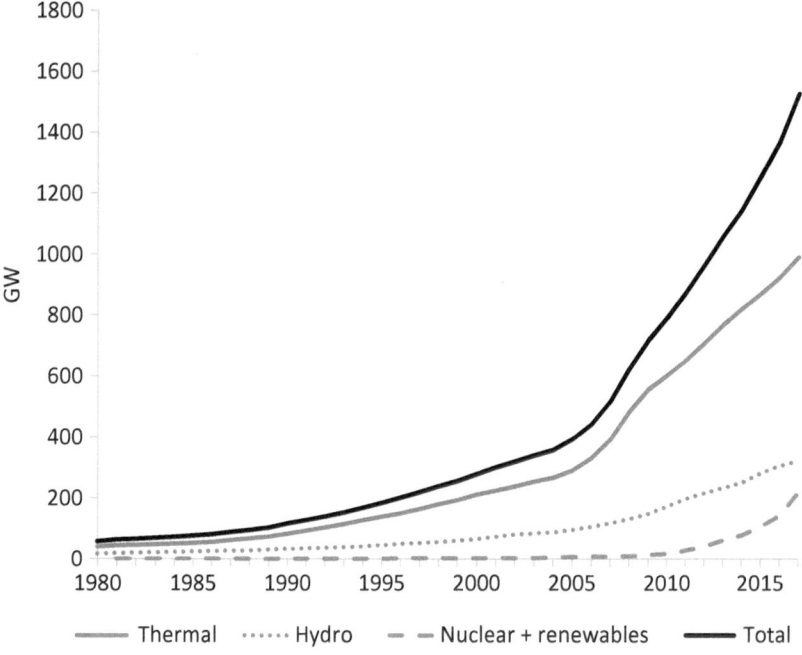

Fig. 2.2 Growth and mix of generating capacity between 1980 and 2015 in GW [11–13]

allowed the government to set electricity tariffs at low levels, especially for energy-intensive industries and households [14]. The government's strategy for energy supply security was directed mainly at raising supply rather than constraining demand and promoting efficiency [4]. At the same time, inadequate attention was paid to the environmental consequences of energy production, transformation and end-use, notably air and water pollution, and greenhouse gas emissions [15, 16].

Since the beginning of the twenty-first century, the government's emphasis has steadily shifted to constraining the rise in energy demand and greenhouse gases, and reducing pollution from the energy sector. This change of priorities arose, in the first instance, from widespread blackouts that started to occur in 2003 as economic growth soared. This led to a surge of construction that boosted installed capacity to 1200 GW by 2013 and a vigorous campaign to reduce national energy intensity. Since 2005, the environment has become an increasingly important topic of public debate and of official pronouncements in China, as both global climate change and domestic environmental degradation became seen as threats to national security and societal well-being [17]. In response, and also in support of industrial policy, the government has promoted the deployment of non-fossil fuel sources of electricity, such as nuclear power, wind and solar PV, as well as sustaining its long-term programme of large-scale hydroelectricity (Fig. 2.2). Nevertheless, domestic coal remains the cheapest fuel and national energy policy continues to be plagued by the tension between energy supply security and environmental protection.

STRUCTURAL EVOLUTION

During the 1980s the government corporatised much of the energy sector, by removing industrial functions from ministries and placing them in newly created corporations. However, the electrical power industry remained under direct ministerial control, despite frequent changes in the structure and name of the relevant ministry [18]. During the subsequent decade, the economic reform agenda was heavily influenced by international organisations such as the World Bank. At the same time, the leadership was concerned with the high cost of government and the large scale of financial losses being incurred by many SOEs, including those in energy production and transformation. Although the economy and the consequent demand for energy continued to grow, the rate of growth declined in the late 1990s as a result of the Asian Financial Crisis.

It was within the context of nationwide industrial reform, of new ideas from abroad and of specific challenges in the energy sector, that the reform of the energy sector and of other industrial sectors was launched in 1998. Over the next five years, the energy sector was transformed through corporatisation, restructuring, commercialisation, listing and the transfer of many regulatory functions from enterprises to government.

The reform of the power industry at this time was much more limited than that experienced by the coal, oil and gas industries. In 1997, the assets of the Ministry of Electrical Power were transferred to the newly created State Power Corporation of China (SPCC), which, as a result, owned most of the transmission and distribution infrastructure and about 50 per cent of the nation's generation capacity. The rest of the assets continued to be owned by a wide variety of SOEs linked to different levels of government, mainly provincial and municipal [10]. The decision not to break up the SPCC at this time appears to have received the support of both ex-Premier Li Peng and the then Premier Zhu Rongji on account of their belief that centralised control of the power sector was necessary. Nevertheless, the government charged the SPCC with reforming the power sector [19].

It soon became apparent that the new structure was deeply unsatisfactory and that the SPCC was making no effort to reform the industry. Intense debates took place involving the highest levels of government and with senior officials publishing their arguments for further reform in the press [18]. This culminated in the decision, at the end of 2002, to dismantle the SPCC in order to separate generation from transmission and distribution, and to reduce the concentration of ownership of power generating capacity. The generating assets of the SPCC were unbundled from the grid and, together with those of the pre-existing Huaneng Group, were assigned to five large corporations whose main business was to be power generation: the Huaneng, Datang, Huadian, Guodian and China Power Investment corporations. The distribution of generating assets to the five new companies was carried out in such a way as to ensure that no single company owned more than 20 per cent of the generating capacity in one of the planned regional power markets. Subsidiaries of these corporations have since been listed on stock exchanges. Although these five companies nominally lay at arm's length from the government, they retained close connections to the political elite [20].

At the same time, the transmission and distribution assets of the SPCC were divided between two new companies. The State Grid Corporation retained the majority of the regional grids in the country, as well as the inter-regional transmission lines, on the basis that central control of the grid was essential. The Southern China Power Grid Company had already been created as a subsidiary of SPCC in order to support the national policy of transmitting electricity from Yunnan to Guangdong, a part of the Great Western Development Strategy that was launched in the year 2000 [18]. It took over the assets in the far south of the country. The two new grid companies were required to progressively sell off most of their generating capacity.

In 2003, the government created what initially appeared to be an entirely new type of regulatory agency, the State Electricity Regulatory Commission (SERC). In principle, SERC was modelled after the classic 'independent regulator'. It reported directly to the State Council and was charged with wide-ranging responsibilities relating to both strategy and regulation. In particular, it was to provide proposals for the development of power markets and the further reform of the wider power sector. However, its power over strategy, investment and economic regulation was deliberately constrained to drafting proposals, as ultimate authority for these matters remained with the NDRC [21]. This anomaly was eventually addressed in 2013 when SERC was formally incorporated into the NEA, thus removing the charade of independent regulation.

Few changes were made to the structure of tariffs. The wholesale tariff and annual operating hours for each plant continued to be set at the provincial level. A number of inconclusive experiments of wholesale competition through power pools were carried out, but no steps were taken towards the systematic introduction of competition in power generation [19]. In 2003, the government proposed further reforms to the system for electricity pricing that would lead to three separate sets of tariffs, for generation (with both capacity and energy components), for transmission and distribution, and for retail, with the eventual separation of transmission and distribution tariffs. But a decade later, these proposals had not yet been implemented.

The government largely suspended the process of reforming the energy sector between 2004 and 2010, mainly because of the shortage in domestic energy supply that was experienced from 2003, as economic growth soared. In the case of the electrical power sector, the government's reluctance to press on with reform could also be attributed to external events. The years 2000 to 2005 saw severe blackouts and politically unacceptable price volatility in a number of liberalised power markets, such as in the USA, Canada, the UK, Scandinavia and Italy. China's government took from these experiences the lesson that electricity sector liberalisation was fraught with risks, and that such risks would be exacerbated in an environment with weak regulatory and legal systems [20, 21].

The slowdown in the economy since 2011 provided the opportunity for the State Council to revitalise its reform efforts, which had been suspended since 2003, by issuing *Document Number 9* in March 2015. The new round of reform is characterised as 'controlling the middle and deregulating the two ends'. This meant that while competition would be gradually introduced into the upstream (generation) and downstream (retail) segments, the midstream (transmission and distribution) would remain regulated.

POLICY INSTRUMENTS

Under the centrally planned economy of Mao, the SPC and its subordinate agencies at local level drew up and oversaw the implementation of plans that covered investment in electrical power infrastructure as well as the production and consumption of electricity. Under Soviet guidance, a large share of the investment in generation during the 1950s was directed at the construction of large-scale hydroelectric dams with the primary aim of supporting heavy industry [10]. Each industrial plant and work unit was allocated a certain amount of electricity supply each year. The chaos that accompanied the Great Leap Forward (1958–61) and the Cultural Revolution (1966–76) undermined these planning processes, and local governments started to take more initiative by finding ways to incentivise investment in power generation. As for all commodities, electricity prices for producers and consumers were set by central government and bore no relationship to the balance between supply and demand. This combination of policies led to constant power shortages from the 1950s to the 1970s.

Deng Xiaoping's industrialisation programme required a sustained increase in electricity supply. Whilst the production of coal grew rapidly, investment in power generation lagged behind the rising demand, leading to continued blackouts [22]. The gradual liberalisation of coal prices in the early 1980s, together with a requirement that new power plants had to repay central government loans with interest, greatly dampened the economic incentives of provincial governments to invest in new thermal capacity. In response, and in line with reforms in other sectors, the government delegated more power for planning and financing electricity infrastructure to provincial governments. It also introduced a two-tier tariff mechanism for power plants, which allowed them to sell at market prices output that exceeded the plan, and allowed end-user tariffs to rise. In addition, new generating plants received higher tariffs than older plants [10]. Consistent with the more open approach to international engagement, the government accelerated a trend of attracting foreign funding that had begun in the late 1980s [23]. These policies lifted total generating capacity from 66 GW in 1980 to 357 GW in 2002 (Fig. 2.2).

The 1990s saw a new emphasis on environmental protection in China. Combined with the long-term need for energy, this led to a more sustained effort by the central government to promote investment in wind and solar energy. In 1994, the Ministry of Electrical Power published the country's first strategic plan for wind energy, which set a goal of 1000 MW of capacity to be installed by 2000. This aspiration was undermined by a combination of the high cost of imported equipment, poor coordination between government agencies, a shortage of financing and the reluctance of the local grid companies to purchase the electricity generated [24]. As a result, only 350 GW had been installed by 2000. High costs also constrained the deployment of solar PV in the 1990s.

Regardless of these policy adjustments, government at central and provincial levels remained firmly in charge of electricity tariffs, for both generators and end-users. Moreover, decisions on dispatch continued to bear no relation to marginal cost. Rather, local governments would approve an annual number of hours to each generating plant, in many cases in direct contradiction of central government directives to preferentially dispatch efficient and clean capacity. No tariff existed for transmission and distribution. Rather, the revenue of the grid companies derived from the difference between the tariffs for generators and end-users. The end-user tariffs continued to be set according to 'catalogues' in which the tariffs varied between different categories and scales of consumer, and between different provinces and municipalities. These tariffs could be supplemented by additional fees. Consistent with long-standing political priorities, the tariffs for heavy industry, agriculture and households were held well below marginal cost [21].

The electricity supply crisis of the early 2000s led to a succession of measures to curb energy consumption across the economy, with particular emphasis on heavy industry including the power sector. The overall objective as stated in the 11th Five-Year Plan was to reduce national energy intensity by 20 per cent between 2005 and 2010 and by a further 16 per cent by 2015, and to reduce the intensity of carbon dioxide emissions by 40–45 per cent between 2005 and 2020.

Measures specific to coal-fired power generation were wide-ranging and had the aim of reducing average net coal consumption from 392 gce/kWh in 2000 to 320 gce/kWh in 2020, as well as reducing air pollution [25]. They included:

- Banning the construction of plants with a capacity of less than 135 MW, announced in 2002;
- Decommissioning plants below 100 MW capacity and replacing small plants with large ones, announced in 2007;
- Prioritising the construction of plants of 600 MW capacity or larger, and the deployment of supercritical and ultra-supercritical technologies;
- Upgrading and retrofitting of older plants that were not closed, mandated in 2012; and
- Building more combined heat and power capacity.

The policy instruments deployed were mainly administrative in nature, for example, through the centralised approval process for investment and through energy efficiency benchmarking. Financial support was provided through compensation for plant closure and loans for new capacity that met the technological requirements. In parallel with these measures, the government has continued the policies initiated by the Great Western Development Strategy by encouraging the siting of new power generating capacity in the western regions of the country. This addresses three objectives: taking advantage of the geographic distribution of primary energy resources (coal, hydro and wind), creating employment in these regions and reducing air pollution from coal-fired plants in the eastern regions.

These policies met with a high degree of success. By the end of 2015 more than 100 GW of small-scale plants had been closed, a total of 219 GW of supercritical and 155 GW of ultra-supercritical plants had been commissioned, and average net coal consumption had declined to 315 gce/kWh [26]. This compares to an average of about 400 gce/kWh in the USA [27] and marks a substantial achievement on the part of China.

This success has been undermined by a separate policy decision in 2013 to relax the need for central government approval for many types of infrastructure project, including thermal power plants. This led to a surge of construction approved by provincial governments that brought 170 GW of coal-fired capacity online between 2012 and 2015, just as annual demand growth was slowing from 12 per cent in 2011 to 0.5 per cent in 2015 [28]. As a result, the average load factor of thermal plants declined from 62 per cent in 2011 to less than 45 per cent in 2015, which would have negatively affected average thermal efficiencies. By this time, a further 200 GW of coal-fired capacity was under construction and permits had been issued for an additional 30 GW [28]. In response, the central government took back the approval process by issuing instructions to provincial governments to delay projects which had not broken ground, and to stop approving new projects unless there was a clear need. At the same time, the NEA issued a further list of some 70 GW of plants to be decommissioned by 2020 [26].

The period from 2003 also saw a sustained effort by the central government to promote renewable energy. The motivations included boosting total electricity supply, increasing the share of clean energy, encouraging technological development and exports, and supporting local development and employment. The Renewable Energy Law of 2005 and supporting policies such as feed-in tariffs and generous loans provided incentives for actors along the full supply chains for wind energy and solar PV. As a result of these and other measures, China quickly became a world leader in the manufacturing of renewable energy equipment [29, 30]. In addition, installed capacity of wind power and solar PV reached 169 GW and 77 GWp, respectively, by the end of 2016 (see Chap. 4 of this book). The capacity of nuclear power also began to rise during this period to 33 GW.

MARKET REFORMS

As part of the energy sector reform programme, a number of inconclusive electricity market experiments were carried out between 1999 and 2005 but were abandoned as the power supply crisis deepened.

The first trials of the new markets were held in Northeast China and East China. The Northeast China power market was put into monthly bidding simulation in January 2004. It initially adopted a one-part price model with 15 per cent of total electricity bid into the market. Following the recommendation of NDRC, the market changed to a two-part price model with all electricity bid into the regional power market. At the beginning, only those generators with a capacity of 100 MW or above were allowed to participate in the pool. During the simulation period, only the bidding system was put into operation, and there was no actual settlement. The East China power market was put into monthly bidding simulation in May 2004, again without actual dispatch and settlement.

Both these pilot markets took the form of a mandatory pool with a single buyer. Bidding to the pool was compulsory for qualified generators, which, in the case of East China, covered coal-fired plants with capacities of 100 MW or greater. The grid company was the single buyer. Trading arrangements were dominated by contract trade and supplemented with trading in the spot market. The trading types included yearly contracts, monthly bidding contracts, day-ahead bidding and real-time balancing. Monthly bidding and day-ahead bidding were operated in the regional trading centre with all the coal-fired units with a capacity of 100 MW or above participating. The provincial dispatching centre was responsible for scheduling the implementation of the annual contracts and for real-time balancing to control the provincial power system.

Further trials were launched in South China in 2005. Unlike the pilot programmes in Northeast and East China, this programme had the intention to stimulate a greater degree of competition. Two characteristics distinguished it from the earlier pilot programmes. Firstly, it engaged not only multiple sellers, but also multiple buyers in the market. The programme required grid companies from four provinces (Guangdong, Guangxi, Yunnan and Guizhou) to participate in the market and these grid companies competed with each other for power purchase. Secondly, the programme separated the dispatch function from the market operator.

The development of these pilot regional markets faced a number of challenges. The varying levels of economic development in different provinces in the same region made it difficult to implement a unified pricing system because the poorer provinces were not able to afford a higher price. Allegations emerged that grid companies were favouring their own generators. The weakness of inter-provincial transmission capacity led to grid congestion. Finally, the growing shortages of power and the threat of rising tariffs rendered these pilot markets irrelevant and the trials were abandoned.

The slowdown in the economy since 2011 has provided the opportunity for the State Council to revitalise its reform efforts, which had been suspended since 2003, by issuing *Document Number 9* in March 2015. The new round of reform was characterised as 'controlling the middle and deregulating the two ends'. This meant that while competition would be gradually introduced into the upstream (generation) and downstream (retail) segments, the midstream (transmission and distribution) would remain regulated [31].

The central element of these reforms was to set transmission and distribution tariffs on the basis of 'allowable cost plus reasonable profit'. Pilot transmission and distribution tariff reforms were implemented in more than 20 provincial grid companies, among which, pilot reforms scheduled to be launched in 14 provincial grid companies in 2017 were brought forward to September 2016.

A second key element of the reform was to open electricity retail to competition. This should progressively allow companies other than the main SOEs to gain access to the electricity retail market and end-users to choose their retailers. The first step of the retail reform was to restart an experiment that was abandoned several years ago to allow large power consumers to purchase electricity directly from generating companies. However, such direct power transactions face many challenges, including the level of transmission and distribution tariffs, the eligibility criteria for generators and the ongoing monopoly position of the grid companies.

A further element of the reform has been to establish relatively independent electricity trading bodies. As of late 2016, there has been no public announcement of any intention either to break up the State Grid Corporation along geographic lines or to separate transmission from distribution.

CONCLUSIONS

The key political and economic importance of China's power sector has meant that change to the governance arrangements has been generally cautious and incremental. As a result, the government remains deeply involved in many aspects of the sector from enterprise ownership to tariff setting and has allowed market forces to play only a limited role. Enterprises owned at central or local government levels dominate the sector along the supply chain and continue to receive generous financial and political support from the state. As a consequence, and because of the continued uncertainty relating to sector reform, foreign direct investment in power generation has remained at a low level since the Asian Financial Crisis of 1997, despite power generation being open to foreign investment. The current excess of generating capacity only exacerbates this situation.

Nevertheless, the sector has achieved remarkable success in many respects: boosting total generating capacity and supply to the level where a surplus exists; deploying advanced technologies in thermal, renewable and nuclear generation, as well as transmission; and building a major export industry. These achievements have come at a huge cost in terms of financial resources and the environment (see Chap. 5 of this book).

The recent round of sector reforms launched since 2015 marks a further incremental step in enhancing the role of market forces that will be accompanied by the introduction of a nationwide market for carbon emissions (announced for 2017 but now delayed). Whilst these initiatives will probably yield some economic and environmental benefits, the gains are likely to be marginal for as long as SOEs continue to enjoy soft budgetary constraints and access to generous finance, and local governments retain influence over investment and dispatch decisions. As in earlier periods of economic reform, the more economically advanced and open provinces in the east of the country are likely to make faster progress in pursuing these electricity and emission market reforms than inland provinces.

REFERENCES

1. Lieberthal K G & Oksenberg M (1988) *Policy Making in China. Leaders, Structures and Processes* (Princeton: Princeton University Press).
2. Grunberg N (2016) 'Revisiting Fragmented Authoritarianism in China's Central Energy Administration', in Brodsgaard K E (ed.) *Chinese Politics as Fragmented Authoritarianism. Earthquakes, Energy and Environment* (Abingdon: Routledge).
3. Davidson M R, Kahrl F & Karplus V (2016) 'Towards a Political Economy Framework for Wind Power. Does China Break the Mould?', Working Paper 2016/32, United Nations University, World Institute for Development Economics Research.
4. Andrews-Speed P (2012) *The Governance of Energy in China: Transition to a Low-Carbon Economy* (London: Palgrave Macmillan).
5. Downs E S (2008) 'Business Interest Groups in Chinese Politics: The Case of the Oil Companies' in Li C (ed.) *China's Changing Political Landscape. Prospects for Democracy* (Washington, DC: Brookings Institution Press).
6. Dorian J P (1994) *Minerals, Energy and Economic Development in China* (Oxford: Clarendon Press).
7. Pan J, Peng W & Others (2006) '*Rural Electrification in China, 1950–2004*, Program on Energy and Sustainable Development', Working Paper No. 60, Stanford University.
8. Yang M & Yu X (1996) 'China's Power Management', *Energy Policy*, 8, 735–757.
9. U.S. Energy Information Administration (2012) *Annual Energy Review 2012*, https://www.eia.gov/totalenergy/data/annual/#electricity, date accessed 6 June 2017.
10. Xu Y-C (2002) *Powering China. Reforming the Electrical Power Sector in China* (Aldershot: Ashgate).
11. Thomson E (2003) *The Chinese Coal Industry: An Economic History* (London: RoutledgeCurzon).
12. National Bureau of Statistics (various years) *China Statistical Yearbook* (Beijing: National Bureau of Statistics).
13. BP (2016) *BP Statistical Review of World Energy* (London: BP).
14. Andrews-Speed P (2004) *Energy Policy and Regulation in the People's Republic of China* (London: Kluwer Law International).
15. Economy E (2004) *The River Runs Black. The Environmental Challenge to China's Future* (Ithaca, N.Y.: Cornell University Press).
16. Delang C O (2016) *China's Air Pollution Problems* (Abingdon: Routledge).
17. Nyman J & Zeng J (2016) 'Securitization in Chinese Climate and Energy Politics', *WIRES Climate Change*, 7, 301–313.
18. Xu Y-C (2017) *Sinews of Power. Politics of the State Grid Corporation of China* (Oxford: Oxford University Press).

19. International Energy Agency (2006) *China Power Sector Reforms. Where to Next?* (Paris: OECD/IEA).
20. Yeh E T & Lewis J I (2004) 'State Power and the Logic of Reform in China's Electricity Sector', *Pacific Affairs*, 77, 437–465.
21. Andrews-Speed P (2013) 'Reform Postponed. The Evolution of China's Electricity Markets', in Sioshansi F (ed.), *Evolution of Global Electricity Markets. New Paradigms, New Challenges, New Approaches* (Amsterdam: Elsevier).
22. Johnson T M (1992) 'China's Power Industry, 1980–1990: Price Reform, and Its Effect on Energy Efficiency', *Energy*, 17, 1085-1092.
23. Li B & Dorian J P (1995) 'Change in China's Power Sector', *Energy Policy*, 23, 619–626.
24. Lema A & Ruby K (2007) 'Between Fragmented Authoritarianism and Policy Coordination: Creating a Chinese Market for Wind Energy', *Energy Policy*, 35, 3879–3890.
25. Yuan J, Na C & Yang M (2016) 'Energy Efficiency and Conservation in China's Power Sector: Progress and Prospects', in Su B & Thomson E (eds.) *China's Energy Efficiency and Conservation. Sectoral Analysis* (Singapore: Springer).
26. Myllyvyrta L & Shen X (2016) *Burning Money. How China Could Squander over One Trillion Yuan on Unneeded Coal-Fired Capacity* (Beijing: Greenpeace).
27. U.S. Energy Information Administration (2016) *Electric Power Annual Report 2016*, https://www.eia.gov/electricity/annual/html/epa_08_01.html, date accessed 6 June 2017.
28. Yuan J, Li P, Wang Y, Liu Q, Shen X, Zhang K & Dong L (2016) 'Coal Power Overcapacity and Investment Bubble in China During 2015–2020', *Energy Policy*, 97, 136–144.
29. Lewis J I (2013) *Green Innovation in China. China's Wind Power Industry and the Global Transition to a Low-Carbon Economy* (New York: Columbia University Press).
30. Matthews J A & Tan H (2015) *China's Renewable Energy Revolution* (Basingstoke: Palgrave Macmillan).
31. National Development and Reform Commission (2015) *Interpretation of Power Sector Reform* (Beijing: People's Publishing House) (in Chinese).

CHAPTER 3

Mapping China's Power Sector Under Market Reform

Xiying Liu

Abstract China has the world's largest installed electricity sector: generation capacity reached 1.6 TW in 2016. A new round of electricity market reform started in 2015, aiming to improve resource allocation efficiency, and ensure that demand is met in a safe, clean and reliable way. This chapter provides a structured map of China's current power sector, including major stakeholders, their roles, and how they interact with each other. How they will be affected by the current reforms is also discussed. This chapter also explains the current electricity tariff system and how it might change in the coming decade under the reform proposals. The chapter ends by highlighting challenges and opportunities.

Keywords Electricity stakeholders • *Document number 9* • Market reform • Transmission tariff • On-grid tariff

X. Liu (✉)
Independent Energy Economist and Policy Analyst, Zug, Switzerland

© The Author(s) 2018
L. Lester, M. Thomas (eds.), *China's Electricity Sector*,
https://doi.org/10.1007/978-981-10-8192-7_3

Summary

In 2014, China's President Xi Jinping declared that the country's economic development had entered a 'new normal' after three decades of double-digit growth. It is expected that Gross Domestic Product (GDP) growth will continue to slow, and issues of structural industrial overcapacity will become more challenging. Electricity demand growth has also dropped significantly, offering an opportunity for market reform in the electrical power sector by releasing the pressure on supply security and tariff increases. In 2015, China started a new round of electricity market reform and issued a key document with the title *Central Committee of the Communist Party of China and State Council—Further Strengthening the Institutional Reform of the Electrical Power Industry* (also called *Document No. 9*). This document lists the key areas for further reform including the electricity pricing mechanism, the power trading market, the opening up of retail markets and increasing the share of the deregulated electricity market.

Driven by the central government, this new round of reform has moved forward fast. Remarkable progress has been made in introducing new market participants and in changing the rules of electricity pricing and power exchange. Reform of the electricity pricing mechanism has affected not only transmission and distribution tariffs but also power trading between qualified market participants. An audit of transmission and distribution costs was carried out nationwide to establish a reasonable, transparent and regulated tariff system that can support high levels of more effective power trading in the market. In March 2017, the National Development and Reform Commission (NDRC) and the National Energy Administration (NEA) jointly released a notice to encourage market players to participate more in the deregulated market and decide prices on their own. After decades of implementing dispatch plans designed to match demand and supply, this change shows the intention and determination of policymakers in the central government for further market reform.

Policy and institutional reform always leads to compromises. Among all the market participants, the grid companies still hold the strongest position in several key areas. First, they own most of the power exchange centres. Although generation companies are encouraged to trade power with other buyers, trades still take place through these centres. Second, grid companies remain the only qualified settlement service provider in the market: generation companies may not trade with grid companies in the deregulated market but can only receive payment for any trades through the grid companies. By holding these settlement rights, grid companies are well placed to maintain liquidity.

Administrative intervention by local government is still common in provincial power markets. Local governments actively connect generation companies with large consumers to facilitate direct power trading, and, as they try to support local economic development with lower energy costs, tend to press for electricity prices that favour large consumers. Power generation companies will certainly need to improve their negotiation capabilities when facing buyers and local governments in the future.

Some retail companies have profited through power trading, yet most of them have not entered the business at all. This is partially down to unclear rules in the retail market, meaning that many companies currently hold a wait-and-see attitude. Another important reason is that they need to build more innovative business models instead of just competing on price. For private and foreign investors in the retail market, the best opportunities for when they compete with state-owned enterprises (SOEs) lie in innovative technologies and advanced energy management services.

Key Insights

- In 2015, China started a new round of market reform in its electrical power sector. It aims to carry forward reform in key areas including electricity pricing mechanisms, power exchange centres, retail markets and dispatch plans based on electricity demand and supply.
- Reform has made substantial progress in auditing and examining transmission and distribution (T&D) tariffs, establishing power exchange centres and attracting investors to enter the retail market.
- Compromises and limitations are inevitable in such complex and difficult reforms. The participation of local governments actively supports power trading, but can also be seen as interference in the market through administrative methods. More importantly, the institutional design largely favours the grid companies and maintains their market power over other participants.

Key Stakeholders in the Power Sector

The NDRC is the government agency that holds broad responsibility for managing national economic and social development under the Chinese State Council. It manages the power sector through various departments, including:

- The Department of Basic Industry, which is responsible for coordinating energy (including power sector) development plans with national economic and social development plans;
- The Department of Resource Conservation and Environmental Protection, which formulates and implements plans for energy and resource conservation and comprehensive utilisation;
- The Department of Climate Change and the Department of High-Tech Industry; and
- The Department of Price, which focuses on price policy formulation and regulation: it recommends policies and reform plans for price adjustment, and sets and adjusts prices and fees administered by the central government [1].

All the policies and announcements on electricity tariffs are released through the Department of Price. More directly, the NEA formulates and implements development plans and promotes institutional reform in the energy sector. It administers the whole energy sector including the power sector, promoting technological development and energy conservation, leading international energy cooperation and providing recommendations on energy price adjustments. In 2013, when the NEA was established, it assumed the functions of the old State Electricity Regulatory Commission and broadened its regulatory responsibility to other sectors including coal, oil and natural gas.

As a highly regulated industry, many other government agencies also play important roles in the power sector, including the State-owned Assets Supervision and Administration Commission, the Ministry of Industry and Information Technology, and the Ministry of Environmental Protection.

China's previous round of electricity market reforms, which took place in 2002, successfully restructured its vertically monopolistic power industry, and set up two national grid companies (State Grid Corporation of China [State Grid] and China Southern Power Grid [Southern Grid]) and five large generation corporations (China Huaneng Group, China Huadian Corporation, China Guodian Corporation, China Datang Corporation and China Power Investment Corporation). In 2017, the structure of the power sector remained almost the same. In the power generation sector, those five power companies now own around half of China's total installed generation capacity, but have been joined by four other major generators: China Resources Power, Shenhua Guohua Electric Power, State Development & Investment Corporation Power Holdings, and China General Nuclear Power Group. All these companies are state-owned corporations with a diverse generation mix, including thermal, hydro, nuclear and renewable energy. Their power generation plants are located across the national market.

In the T&D sector, State Grid and Southern Grid operate at a national level [2]. Southern Grid covers five provinces and one autonomous region in the south: Guangdong Province, Yunnan Province, Guizhou Province, Hainan Province and Guangxi Zhuang Autonomous Region. The State Grid operates in the rest of the country, accounting for 89 per cent of China's area. Both State Grid and Southern Grid are Fortune Global 500 companies, ranked 2nd and 100th, respectively, in 2017. Their respective 2016 revenues were RMB 2.1 trillion and RMB 476.5 billion. In addition to the domestic market, State Grid also owns and operates overseas assets in the Philippines, Brazil, Portugal, Australia and Italy. Southern Grid has been actively involved in power transactions in Hong Kong and Macau, as well as with countries in the Greater Mekong region, such as Vietnam and Laos.

Besides these two giant grid companies, there is an independent provincial grid operator called the Inner Mongolia Power Group (also known as the Inner Mongolia Western Grid) [3], which covers the middle and western part of Inner Mongolia Autonomous Region. It owns more than half of the installed generation capacity in Inner Mongolia and runs independently from both State Grid and Southern Grid. Inner Mongolia has large amounts of coal and natural gas, and sufficient renewable energy like wind and solar, making it a designated strategic energy and resource supporting base for national development [4]. Unfortunately, the Inner Mongolia Western Grid faces challenges in the form of electricity transmission constraints. Until it succeeds in reaching an agreement with the State Grid, and is able to transmit electricity to other regions via the transmission lines of State Grid, the Inner Mongolia Western Grid has to use almost all electricity locally. So far, it has only one cross-region transmission line, which connects to the Shaanxi Regional Electric Power Group Company (a state-owned local power distribution company) and provides electricity to 66 counties in Shaanxi Province. The Inner Mongolia Autonomous Region set out a plan for a new transmission network in its Energy Development 13th Five-year Plan to improve its capacity for cross-region electricity transmission, especially for wind and solar energy. However, implementation still needs some years to be examined. In addition, there are another 15 companies in other provinces/cities that own local distribution networks, provide distribution services and sometimes also own generation capacity.

Document No. 9 also opens the retail market to various investors and companies. Hundreds of new retail companies have registered themselves and some of them now participate in power exchanges, exploring brand new business opportunities in the electricity market. However, the electricity pricing mechanism is still highly regulated and influenced by central and local governments, and both the retail market and power trade rules still need to be improved. Therefore, these retail markets often show immature features and have led to many disputes.

In terms of overall demand, agriculture, industry, the service sector and residential consumers account for 1.8 per cent, 71.1 per cent, 13.4 per cent and 13.6 per cent of total electricity consumption, respectively [5].

ELECTRICITY TARIFF SYSTEM

Despite many years of efforts to push market-oriented reforms, electricity tariffs in China remain under full regulation. This is at least in part because they are often used as policy tools for various macro goals far beyond the electricity sector. China's electricity tariff system is complex, consisting of different types of tariffs for different sectors in the power industry. From the generation sector, through T&D, to end-users, the three most important types of tariffs are on-grid tariffs, T&D tariffs and retail prices. The NDRC is the government agency that determines the prices of regulated goods.

On-Grid Tariffs

On-grid tariffs are set for power generation companies to sell electricity to grid companies. Depending on fuel, different generators face different prices. For example, renewable energy companies, such as solar photovoltaic (PV) and wind energy generators have higher on-grid tariffs (meaning that they can sell their electricity at a higher price) than fossil fuel power generators and nuclear power plants. Hydropower plants usually have the lowest on-grid tariffs.

Coal-Fired Power Generators

In 2004, the NDRC set up a benchmark on-grid tariff system for coal-fired generators. Based on the average generation costs in each province/autonomous region, the NDRC estimates and releases the on-grid tariffs for newly built coal-fired power generators in that province/autonomous region. The calculation takes many factors into consideration, including fuel prices, total investment, annual generation hours and interest rates on outstanding loans.

Nuclear Power Plants
Traditionally, each nuclear power plant in China had its own on-grid tariff. It was examined and approved by the NDRC, based mainly on individual investment and cost structures. In 2013, the NDRC set up a national benchmark on-grid tariff for nuclear power plants built after 1 January 2013, at RMB 0.43/kWh, based on the average cost and electricity supply-demand situation. Local benchmark on-grid thermal power tariffs (including desulfurization and denitrification price mark-ups) are also used as adjustment factors.

When local on-grid tariffs for coal-fired power plants are higher than RMB 0.43/kWh, nuclear power plants can only sell their electricity to grid companies at the national nuclear benchmark tariff; if the provincial benchmark for on-grid thermal power tariffs where they are located is lower than this, they may only charge up to the thermal tariff [6]. This policy has affected the allocation strategy of new nuclear power plants and encouraged them to move to provinces with higher potential power shortages given that on-grid thermal power tariffs tend to be lower in provinces with sufficient power supply and higher in provinces with shortages or higher potential for shortages.

Hydropower Plants
In the beginning of 2014, the NDRC released the *Completion of Hydropower On-Grid Tariff Pricing Mechanism* by which both provincial and cross-province/cross-region hydropower on-grid tariffs were decided for hydropower plants built after 1 February 2014. In principle, hydropower on-grid tariffs are based on the average electricity purchasing price of the grid company in the province, on the supply-demand relationship and on the development costs of hydropower. In provinces where hydropower accounts for a high share of the electricity mix, the hydropower on-grid tariff may also integrate seasonal and Time-of-Use prices.

Cross-province and cross-region hydropower trades involve a more complex pricing framework consisting of on-grid tariffs in the selling province, transmission tariffs and landing prices in the receiving province.

Landing prices are used as the starting point in price determination, and the calculations work backwards to decide the actual on-grid tariff. The on-grid tariff that hydropower plants actually receive is then the negotiated price less the transmission tariff.

Transmission tariffs are fully regulated. The NDRC sets the cross-province electricity transmission tariffs; the NEA examines the cross-region transmission tariffs and reports to the NDRC for approval. The landing prices can be negotiated by hydropower plants and buyers in the receiving province based on the average on-grid tariffs in receiving provinces. Because hydropower is mainly transmitted from the less-developed western region to the more-developed eastern region, hydropower plants in selling provinces tend to have low benchmark on-grid tariffs and can, therefore, receive higher on-grid tariffs through this pricing mechanism [7]. Some selected regional on-grid tariffs are shown in Table 3.1.

Renewable Power Plants
After years of administratively approved pricing and bidding price mechanisms, the NDRC put a benchmark feed-in tariff framework for onshore wind power in place in 2009. This consisted of a four-tier price system ranging from RMB 0.51–0.61/kWh based on local wind resources (see Table 3.2) [8]. By 2018, the feed-in incentives will be gradually reduced to RMB 0.44–0.58/kWh. Currently, the benchmark feed-in tariffs for offshore and inter-tidal wind farms (non-bidding projects) are set at RMB 0.85/kWh and RMB 0.75/kWh, respectively [9].

Based on the different solar resources and construction costs across the country, the NDRC also set up a three-tier feed-in tariff system for solar PV farms in 2013 (see Table 3.3) [10]. As the cost of solar PV reduces, the feed-in incentives are gradually being lowered from RMB 0.90–1.00/kWh to RMB 0.65–0.85/kWh (see Table 3.3) [9]. Starting from 2010, the feed-in tariff for biomass power has been set at RMB 0.75/kWh for projects whose prices are not decided by auction [11].

T&D Tariffs

Grid companies receive T&D tariffs less on-grid tariffs, government funds and additional charges from end-user tariffs. Because on-grid tariffs, government funds, additional charges and end-user tariffs are all regulated, T&D tariffs are also a product of regulation, but without an independent or clear pricing mechanism. Policymakers have targeted the T&D pricing mechanism for reform in the new round of reform.

Table 3.1 Average on-grid tariffs for power generation companies, unit: RMB/MWh, %

Province	Coal-fired power plants			Gas-fired power plants			Hydropower		
	2015	2014	Growth rate	2015	2014	Growth rate	2015	2014	Growth rate
Nation average	384.2	400.89	−4.16	789.82	750.78	5.2	286.93	297.76	−3.64
Beijing	467.81	514.61	−9.09	752.87	650	15.83			
Tianjin	418.65	430.3	−2.71	982.88	650	51.21			
Hebei	404.47	425.39	−4.92						
Shanxi	329.2	384.42	−14.36				359.75	369.64	−2.68
Shandong	421.06	456.79	−7.82						
Inner Mongolia East	300.34	283.68	5.87				318.33	318.24	0.03
Inner Mongolia West	284.59	318.58	−10.59						
Liaoning	390.49	363.98	7.28				374	692.75	−46.01
Jilin	370.57	321.39	15.3				426.73	392.07	8.84
Heilongjiang	389.52	401.67	−3.02				516.41	392.07	31.71
Shaanxi	378.85	380.59	−0.46				335.49	321.67	4.3
Gansu	295.92	324.79	−8.89				239.5	237.25	0.95
Ningxia	264.42	280.82	−5.84				263.86	265.82	−0.74
Qinghai	335.88	348.36	−3.58				281.49	256.31	9.82
Xinjiang	247.34	253.9	−2.58				248.84	245.19	1.49
Shanghai	441.05	459.87	−4.09	826.17	810.03	1.99			
Zhejiang	472.67	482.57	−2.35	992.51	956.4	3.78	552.31	565.98	−2.42
Jiangsu	414.95	419.21	−1.02	736.83	853.54	−13.67			
Anhui	413.19	434.69	−4.95				431.58	373	15.71
Fujian	392.86	428.45	−8.31	1260.89	588.32	114.32	382.41	301.43	26.87
Hubei	442.74	466.85	−5.16	880.12	962.83	−8.59	267.42	291.89	−8.38
Henan	421.26	454.76	−7.37	786.53	609.04	29.14	340.35	357.39	−4.77
Hunan	473.66	489.17	−3.17				361.27	372.65	−3.05

(continued)

Table 3.1 (continued)

Province	Coal-fired power plants			Gas-fired power plants			Hydropower		
	2015	2014	Growth rate	2015	2014	Growth rate	2015	2014	Growth rate
Jiangxi	443.73	469.39	−5.47				281.98	281.81	0.06
Sichuan	464.36	472.43	−1.17	505.37	814.98	−37.99	298.14	311.38	−4.25
Chongqing	433.32	445.77	−2.79				315.1	314.65	0.14
Guangdong	504.76	536.36	−5.89	649.28	589.19	10.2	273.97	272.65	0.48
Guangxi	488.87	488.42	0.09	1283.56			276.92	284.06	−2.51
Yunnan	462.91	356.66	29.79				252.94	261.9	−3.42
Guizhou	365.1	380.69	−4.1				296.34	291.01	1.83
Hainan	459.32	471.16	−2.51	628.85	445.61	41.12	402.22	394.75	1.89

Notice: Average on-grid tariff (including tax) = revenue from selling electricity/on-grid electricity × 1.17

Source: National Energy Administration, 2016, Supervision Notification of National Power Electricity Price Situation 2015, http://zfxxgk.nea.gov.cn/auto92/201611/t20161101_2312.htm, data accessed 10 October 2017

Table 3.2   Benchmark feed-in tariffs for onshore wind power generation, unit: RMB/kWh (including tax)

Resource region	Benchmark feed-in tariffs for onshore wind projects built after 1 January 2018	Areas included in the region
Category I	0.4	Inner Mongolia Autonomous Region (except for Chifeng, Tongliao, Hinggan League and Hulun Buir); and Urumqi, Ili Kazak autonomous prefecture, Karamay and Shihezi in Xinjiang Autonomous Region
Category II	0.45	Zhangjiakou and Chengde in Hebei province; Chifeng, Tongliao, Hinggan League and Hulun Buir in Inner Mongolia Autonomous Region; Jiayuguan and Jiuquan in Gansu province; and Yunnan province
Category III	0.49	Baicheng and Songyuan in Jilin province; Jixi, Shuangyashan, Qitaihe, Suihua, Yichun and Greater Khingan area in Heilongjiang province; Gansu province except for Jiayuguan and Jiuquan; Xinjiang Autonomous Region except for Urumqi, Ili Kazak autonomous prefecture, Karamay and Shihezi; and Ningxia Autonomous Region
Category IV	0.57	All other regions

Source: NDRC (2016), http://www.ndrc.gov.cn/zwfwzx/zfdj/jggg/201612/t20161228_833062. html, date accessed 20 July 2017

Retail Prices

Retail Price Structure

For end-users, retail prices are designated for three types of consumers:

- Residential;
- Agricultural and industrial; and
- Commercial and other consumers [12].

Retail prices are calculated based on the following equation:

$$\text{end-user tariff} = \text{on-grid tariff} + \text{T\&D tariff} + \text{line loss} + \\ \text{government funds and additional charges}$$

Table 3.3   Benchmark feed-in tariffs for solar PV power generation, unit: RMB/ kWh (including tax)

Resource region	Benchmark feed-in tariffs for onshore wind projects built after 1 January 2017	Areas included in the region
Category I	0.65	Ningxia; Haisi zhou in Qinghai province; Jiayuguan, Wuwei, Zhangye, Jiuquan, Dunhuang and Jinchang in Gansu province; Kumul, Tacheng, Altay and Karamay in Xinjiang; Inner Mongolia Autonomous Region except for Chifeng, Tongliao, Hinggan League and Hulun Buir
Category II	0.75	Beijing, Tianjin, Heilongjiang, Jilin, Liaoning, Sichuan, Yunnan, Chifeng, Tongliao, Hinggan League and Hulun Buir in Inner Mongolia Autonomous Region; Chengdu, Zhangjiakou, Tangshan and Qinhuangdao in Hebei province; Datong, Shuozhou, Xinzhou and Yangquan in Shanxi province; Yulin and Yan'an in Shaanxi province; Qinghai; Gansu; and Xinjiang except for those in Category I
Category III	0.85	All other regions

Source: NDRC (2016), http://www.ndrc.gov.cn/zwfwzx/zfdj/jggg/201612/t20161228_833062. html, date accessed 20 July 2017

Currently, large and complex cross-subsidies exist between different groups of consumers. Differing from most other countries, residential and agricultural customers in China pay lower electricity prices than industrial and commercial consumers. Large industrial and commercial customers subsidise the residential sector. Meanwhile, urban regions subsidise the rural regions (Table 3.1). The tariff framework was designed to support the purchasing power of residents and the agricultural sector. However, the system does not reflect the real cost of electricity and leads to efficiency losses. Given the unclear pricing mechanism in the T&D sector, and complex cross-subsidies in both T&D and the end-user sector, it is also difficult to calculate the value of these cross-subsidies and the efficiency losses caused by them.

The new round of electricity market reform seeks to address these cross-subsidy issues along with the T&D tariff reform. It aims to clarify the cross-subsidy each group of consumers receives and to set up a new T&D tariff system that reflects costs and losses at different voltage levels (Table 3.4).

Table 3.4 Electricity end-user tariffs in Jiangsu Province, January 2016

End-user groups			Energy tariff < 1KV	[1KV, 10KV]	[20 KV, 35 KV]	[35 KV, 110 KV]	>100 KV	≥ 220 KV	Capacity tariff Maximum (RMB/KV·Month)	Transformer capacity (RMB/KVA·month)
1. Residential sector	Tiered pricing	Annual consumption ≤ 2,760kWh	0.5283	0.5183						
		(2,760kWh, 4,000kWh]	0.5783	0.5683						
		> 4,000kWh	0.8283	0.8183						
	Others		0.5483	0.5383						
2. Medium commercial and industrial and other sectors			0.8439	0.8289	0.8229	0.8139				
(1) Restrictive energy-intensive industry			0.9439	0.9289	0.9229	0.9139				
(2) Eliminative energy-intensive industry			1.1439	1.1289	1.1229	1.1139				
3. Large industrial sector				0.6601	0.6541	0.6451	0.6301	0.6151	40	30
(1) Restrictive energy-intensive industry				0.7601	0.7541	0.7451	0.7301	0.7151	40	30
(2) Eliminative energy-intensive industry				0.9601	0.9541	0.9451	0.9301	0.9151	40	30
4. Agricultural sector			0.509	0.499	0.493	0.484				

Source: Jiangsu Province Bureau Document (2016) No1, http://www.jiangyin.gov.cn (home page), data accessed 20 August 2017

Regional Differences in End-User Tariffs

Electricity retail prices show significant differences between provinces for various reasons, including the power generation fuel mix, resource endowment, local demand-supply relationships and T&D capacity. In 2015, Shenzhen had the highest average retail price at RMB 807.88/MWh, and Qinghai had the lowest at RMB 381.37/MWh. Some provinces in the west have relatively low prices stemming from high supply and limited demand; provinces in the northwest and southwest (Qinghai, Ningxia and Inner Mongolia) have slower economic development compared to those in the east. Eastern and coastal provinces usually have much higher electricity retail prices. As a generalisation, western regions also have more resources, including both conventional fossil fuel and hydropower, and renewable energy like wind and solar power. For example, the average retail price (across all types of end-user) is RMB 770.09/MWh in Guangzhou and RMB 760.34/MWh in Shanghai. With lower income levels, residents in the west also tend to pay less for energy compared to those who live in the east. For example, end-user tariffs for Ningxia and Qinghai are RMB 457.24/MWh and RMB 405.80/MWh, respectively, while residents in Shenzhen and Guangzhou pay RMB 714.62/MWh and RMB 654.08/MWh, respectively (Table 3.5).

Table 3.5 Average retail price for grid companies and end-user tariffs for residential consumers, unit: RMB/MWh, %

Region	Average retail price for grid companies			End-user tariffs for residential consumers		
	2015	2014	Growth rate	2015	2014	Growth rate
National average	643.33	647.05	−0.57	548.04	557.48	−1.69
Beijing	777.33	776.09	0.16	495.12	495.68	−0.11
Tianjin	725.5	719.34	0.86	503.25	502.37	0.18
Hebei (north grid)	587.68	596.32	−1.45	514.33	514.91	−0.11
Hebei (south grid)	641.34	664.24	−3.45	524.14	524.74	−0.11
Shanxi	510.26	520.66	−2	485.77	485.95	−0.04
Shandong	697.75	711.82	−1.98	536.71	535.66	0.2
Inner Mongolia East	513.2	556.32	−7.75	507.03	504.03	0.6
Inner Mongolia West	420.77	400.88	4.96	440.49	439.81	0.15
Liaoning	613.06	628.12	−2.4	512.38	511.17	0.24
Jilin	630.54	625.87	0.75	534.76	533.8	0.18

(*continued*)

Table 3.5   (continued)

Region	Average retail price for grid companies			End-user tariffs for residential consumers		
	2015	2014	Growth rate	2015	2014	Growth rate
Heilongjiang	547.14	559.25	−2.17	482.22	480.81	0.29
Shaanxi	554.73	560.45	−1.02	507.07	507.35	−0.06
Gansu	453	461.91	−1.93	526.06	526.4	−0.06
Ningxia	393.74	407.33	−3.34	457.24	456.31	0.2
Qinghai	381.37	384.24	−0.75	405.8	406.99	−0.29
Xinjiang	436.81	444.21	−1.67	533.57	531.62	0.37
Shanghai	760.34	768.83	−1.1	571.23	569.73	0.26
Zhejiang	747.25	753.64	−0.85	556.1	556.62	−0.09
Jiangsu	688.79	693.94	−0.74	517.57	519.63	−0.4
Anhui	676.17	690.14	−2.02	569.24	568.63	0.11
Fujian	644.76	668.68	−3.58	551.49	557.33	−1.05
Hubei	669.6	674.88	−0.78	579.5	585.79	−1.07
Henan	606.83	569.42	6.57	563.22	569.83	−1.16
Hunan	675.06	672.83	0.33	607.02	607.38	−0.06
Jiangxi	711.61	732.71	−2.88	618.23	618.5	−0.04
Sichuan	531.64	549.84	−3.31	523.33	531	−1.44
Chongqing	648.45	643.31	0.8	537.34	538.39	−0.2
Guangdong	697.65	714.83	−2.4	674.93	647.04	4.31
Guangxi	557.29	566.79	−1.68	562.87	461.34	22.01
Yunnan	419.22	444.24	−5.63	471.97	476.13	−0.94
Guizhou	492.88	511.95	−3.72	485.44	484.77	0.14
Hainan	734.24	744.18	−1.34	632.06	633.13	−0.17
Guangzhou	770.09	788.64	−2.35	654.08	654.24	−0.02
Shenzhen	807.88	827.82	−2.41	714.62	715.17	−0.08
State Grid (parent company)	398.93	412.61	−3.32	–	–	–
South Grid (parent company)	437.35	459.42	−4.8	–	–	–

Notice: Average retail tariffs do not include governmental fund and additional charges (incl. tax)

Source: National Energy Administration, 2016, Supervision Notification of National Power Electricity Price Situation 2015, http://zfxxgk.nea.gov.cn/auto92/201611/t20161101_2312.htm, data accessed 10 October 2017

Hindered Transmission Mechanism for Price Signals: The Coal-Electricity Price Linkage Mechanism

Even with the world's largest installed capacity of wind and solar power, China's power generation remains dominated by coal-fired power plants. At the end of June 2017, 65 per cent of China's total installed capacity relied on fossil fuels, mainly coal. These plants accounted for about 75 per cent of China's total power generation during January to June 2017 [13]. The power generation sector accounts for about half of China's total coal demand.

Although the coal and electricity sectors are closely connected, they have different pricing mechanisms. On-grid tariffs faced by power generators are regulated; in contrast, the coal market was liberalised in 1993 when market-oriented pricing mechanisms were first introduced. Coal-fired power plants, therefore, face volatile coal prices, but are unable to pass this volatility through to consumers when they sell electricity to the grid companies. There have been various attempts by policymakers to reduce this pricing disconnect. From 1996 to 2001, the State Planning Commission implemented a policy of 'Guiding Price for Thermal Coal', which set ceilings for any increases in the price of thermal coal each year. This policy introduced a dual track pricing mechanism in coal markets: any increase in thermal coal prices must be within the guiding prices released by the State Development Planning Commission (SDPC) every year, while non-thermal coal prices (for use outside the heat and power sector) were fully deregulated. The aim was to ensure sufficient coal supply while maintaining affordable coal prices for the power generators [14]. Coal buyers and sellers, including all types of coal production companies, could only negotiate settlement prices within these ceilings when signing coal purchase contracts at the Coal Ordering Meeting each year. However, the dual track pricing mechanism could not stop coal companies from selling in the deregulated non-thermal coal market and, therefore, did not ensure sufficient supply of coal for power generation companies.

In 2002, the national guiding price for thermal coal was cancelled and replaced by a 'coordinating price for reference' in order to eliminate direct administrative intervention in the coal market. In 2005, the regulation of non-thermal coal prices was completely terminated. Meanwhile, the Coal Ordering Meetings also stopped. Instead, coal production companies and power generation companies were asked to negotiate mid- and long-term (two to five years) purchase contracts, mainly based on the demand-supply relationships in the market. In 2012, as economic growth slowed, coal prices in the deregulated market dropped as demand fell. Taking this opportunity, policymakers stopped temporary administrative interventions in thermal coal prices; from January 2013 the market was to be the sole price determinant [15]. At this point, coal's dual-track pricing mechanism (a regulated thermal coal price and market-oriented price for all other coal) ended.

The coal-electricity price linkage mechanism (CEPLM) was another important method used to enable coal-fired generators to pass on price fluctuations. It was announced by the NDRC in 2004 through a policy paper called *Opinions on Establishing the Coal-Electricity Price Linkage Mechanism* [16]. Aiming to introduce a market-based pricing mechanism to the electricity industry in a step-wise fashion, the CEPLM initially allocated 30 per cent of any coal price increase to the coal-fired power plants and the rest to retail electricity prices (allowing 70 per cent to be passed on). It created a transmission mechanism for price signals from the coal sector to electricity consumers but it was not automatic or real-time. Changes to retail prices could be made at most once a year and it required a hearing before any adjustments to electricity tariffs for residential consumers were approved. For various reasons, the CEPLM was not strictly followed when coal prices fluctuated. In particular, during periods of rapid coal price increases, cost increases were not passed through for fear of wider macro effects stemming from higher energy costs for manufacturing or residents.

After a few rounds of revision, the NDRC issued the latest version of CEPLM in 2015: the *Notice of Improving Coal-Electricity Price Linkage Mechanism*, which set up a mechanism for 2016–20 [17]. It uses the average price of thermal coal nationwide in 2014 as the coal benchmark price, and calculates corresponding on-grid electricity tariffs as a power benchmark price. When changes in coal prices are between RMB 30–150 per ton, both on-grid tariffs and retail prices are updated based on specified equations. The NDRC is responsible for deciding whether to adjust electricity prices or not, though, again, adjustments happen yearly.

Given a market-oriented coal industry and a regulated power generation industry, a price linkage mechanism that connects the two industries more organically and provides a channel for sending price signals about fuel costs to consumers should improve economic efficiency. However, electricity pricing reform is merely one of many reform goals for policymakers in China. Higher energy prices will affect everything from the global competitiveness of China's industry to living costs for residents. After decades of subsidised electricity bills, it is difficult for policymakers to raise prices, in particular during an economic slowdown when they would rather use electricity prices as a regulatory tool to boost economic growth and gain public support. The challenge facing electricity tariff reform is that it is often sacrificed or postponed to achieve other economic development policy goals.

Capped electricity prices have negative effects on market participants and market structure. When coal-fired power plants can only sell electricity at regulated on-grid tariffs, they lose many of their risk and cost management tools in the face of fluctuating fuel costs. The resulting risk profile makes it very unlikely private or foreign investment enters the market; only state-owned or local-government-owned companies can stay in the power generation market. This is preventing China from building either a dynamic ownership structure or simply improving economic and management efficiency within the generating sector.

In the long run, the conflicts between coal price and electricity price will need to be addressed, either through improved price linkage mechanisms that fully reflect market dynamics in a timely manner, or by merging coal producers and power generators to enable internal hedging through transfer pricing.

A New Round of Electricity Market Reform

For decades, the biggest challenge China faced in its electrical power sector was to provide cheap and sufficient electricity to support its model of rapid economic growth driven by industrialisation and urbanisation. Security of supply was, to a large extent, assured by a power sector dominated by SOEs and fully regulated electricity tariffs. Market reform has, on the whole, proceeded only very slowly because of policymakers' concerns over potential supply shortages and price fluctuations.

However, as China's economic growth has slowed, this situation has started to change. In 2014, Chinese President Xi Jinping raised the 'new normal' theory and said that the whole country should adapt to a new phase in China's GDP growth, as it slows down compared to the previous three decades. The following year, at the 11th meeting of the Leading Group for Financial and Economic Affairs, President Xi raised 'supply-side reform' for the first time. In 2015, Premier Li Keqiang delivered a speech at the opening of China's annual parliamentary meeting and identified structural overcapacity in the industrial sector as a key challenge. He listed major policies to tackle the issue, including reducing the capacities of energy-intensive industries and closing outdated facilities. Aiming to stabilize economic growth at around 7 per cent, structural supply-side reforms moved centre stage.

The slowing down of both economic development and electricity demand growth offered an opportunity for the Chinese government to proceed with electricity market reform because there would be less pressure from concerns over security of supply and tariff increases [18]. During 2000–13, the average annual growth rate of electricity consumption was over 10 per cent [19]; in 2014, it was just 3.8 per cent; in 2015, it was 0.5 per cent [20]. During periods of slower growth in electricity demand, market fluctuations will not necessarily lead to electricity supply shortages or rocketing electricity prices. Therefore, policymakers will be less worried about any market instability that may be introduced by reforms and be more willing to make the attempt.

Reform in the electrical power industry is considered to be an important part of China's overall structural supply-side reforms. Generation facilities with outdated technologies and high levels of emissions will be shut down to reduce both overcapacity and pollution within the electrical power industry. Reforming the electricity pricing mechanism will also help to balance the demand-supply relationship, allowing resources to be allocated more efficiently.

In March 2015, China started a new round of electricity market reform by releasing the *Central Committee of the Communist Party of China and State Council—Further Strengthening the Institutional Reform of the Electrical Power Industry* (also called *Document No. 9*). *Document No. 9* provides clear evidence of the central government's determination to continue its agenda of electricity market reform, and highlighted the key areas which can expect major reform, including electricity tariffs, electricity trading schemes, the deregulation of dispatch plans and the opening up of the distribution and retail sectors [21].

In contrast to how past electricity market reform attempts have progressed during the last few decades, the current reform programme is moving forward fast, driven by the central government. Six months after the release of *Document No. 9*, six supporting documents were jointly released by the NDRC and the NEA [22], each of them focusing on a different key reform field, comprising:

1. Electricity tariff reform in the T&D sector;
2. Building the electricity market;
3. Power exchange centres and operational rules;

4. Deregulation of the electricity dispatch plan;
5. Reform of the retail sector; and
6. Supervising coal-fired self-generation power plants.

Based on the general guidelines and fundamental framework set up by *Document No. 9*, these supporting documents explain in more detail the principles, contents, approaches and institutional arrangements necessary in each field.

This new round of reform has had various consequences for the major stakeholders in China's electrical power sector, affecting their roles and how they interact with each other. For example, it has introduced new players into the market and shut down many old power generation plants; many existing players have been required to adjust their business models according to the new policy. In addition, these reforms have also changed the market rules and regulations, as well as how they will impact the market development and future reform.

CHANGES OF MAJOR STAKEHOLDERS IN THE ELECTRICITY MARKET

In addition to existing stakeholders in the electricity market, such as generation companies, grid companies and consumers, the current reform programme has introduced two new types of participants to the market, namely power exchange centres and retailers. The entrance of new players is expected to change both the map of key stakeholders in the electrical power sector and how they interact with each other.

Power Exchange Centres

With the aim of replacing fully regulated electricity dispatch plans, power exchange centres are being set up to provide a platform for trading. By September 2016, 33 power exchange centres had been launched in China: one in each province except for Hainan and an additional two national power trading centres in Beijing and Guangzhou. The Beijing centre is responsible for cross-province and cross-region electricity market development and operation, while the Guangzhou centre focuses more on implementing the West-East Electricity Transfer Strategy, both of which will also promote market integration and optimise resource allocation nationwide [23].

In one of the supporting documents, *Opinions on the Implementation of Establishment and Normative Operation of Electrical Power Exchange Organisations*, all power trading centres are defined as not-for-profit organisations. Their major responsibilities are to provide normative, open and transparent power trading services as required by the regulations. They should:

- Establish, operate and manage the exchange platforms;
- Provide registration, trade and settlement services; and
- Publish market information to the public.

With these services, policymakers expect to build up a unified, open and competitive electricity market to replace regulated power dispatch plans and electricity tariffs.

The power exchange centres have three choices of ownership structure:

1. An independent enterprise, where grid companies are the relative controlling shareholder, and other parties like generators, retailers and consumers are also shareholders;
2. A subsidiary enterprise of a grid company; and
3. A membership institution.

In practice, a large share of power exchange centres have been set up as subsidiary enterprises of the State Grid. Other grid companies, such as Southern Grid and Guizhou Power Grid Corporation, are the largest shareholders of almost all the other power exchange centres. Separately, Guangdong Power Exchange has the Guangdong Grid Corporation as its controlling shareholder along with other companies and organisations, including generation companies, retail companies and third-party organisations in Guangdong province. The participation of other market players remains at very low level.

The goal behind setting up independent power exchange centres was to provide the opportunity for open and fair trade in the electricity market. However, reality is far from intention. By owning these exchange centres, grid companies are able to retain their key role in the power exchange, particularly in the following three fields:

1. Operating new trading platforms;
2. Participating in power trading by themselves and through their subsidiary retail companies; and
3. Providing the only legal settlement services.

Acting as the settlement agency not only provides grid companies with additional cash flow, but also strengthens their market power over power generation companies, non-grid-owned retail companies and large consumers.

In conclusion, the introduction of new and relatively independent power exchange centres has provided the opportunity for more competitive power trading among various market participants but substantial compromises remain embedded in the system.

New Electricity Retail Companies

The opening up of the electricity retail sector was a major highlight of the current reform programme. It attracted large amounts of attention by opening up the valuable retail market to all types of interested investors. The threshold capital for registering an electricity retail company is RMB 2 million. Companies meeting the minimum registered capital requirement, along with other rules, can register themselves as retail companies in the same way as they can register as any other type of company: no special administrative approval is needed. By the end of May 2017, 960 retail companies had registered in the area operated by the State Grid [24], including a large number of private investors.

In addition to private and foreign investors seeking the opportunity to enter China's power sector, the existing key players in the market (the five largest state-owned power generation companies) have set up their own subsidiary retail companies. Entering the retail business is an effective way of managing risk for generation companies, enabling them to balance out electricity price fluctuations.

The reform of electricity dispatch plans is designed to increase the share of power generation traded through the market. Although generation companies are encouraged to enter the market and explore opportunities and increased profits, most coal-fired power plants are unable to profit in the current half-deregulated market. As soon as the coal price drops, they are required to lower their prices and yet may find it hard to raise prices when the coal price increases. In addition, they also face inaccurate market expectations that reform will definitely lead to lower prices.

According to the supporting document, *The Opinions on the Implementation of Promoting Retail-Side Reform*, there are three categories of retail companies:

1. Subsidiary companies of grids;
2. Companies that own distribution grid assets and have been invested in by non-state owned companies; and
3. Independent retail companies.

All the retail companies so far registered in the power exchange centres are qualified to trade with other registered power generation companies, consumers and other retailers. They are free to choose between bilateral trades and trades through centralised exchanges, and will receive the settlement documents from the exchange centres.

Grid companies are also entitled to set up their own subsidiary retail companies. These companies are subject to government regulation in order to create a fair and competitive environment. However, these regulations need high-level oversight to ensure their proper implementation.

Some local governments are also actively involved in the retail market but with a different goal. Most of them expect to lower local electricity tariffs, especially for large industrial consumers, with the aim of attracting more investment and so boosting the local economy. However, market reform does not necessarily lead to lower tariffs; instead, it should improve efficiency and bring a market-oriented pricing mechanism.

Local governments can influence the new market from two perspectives.

1. They strongly encourage direct trade between generation companies and large consumers/retailers. Sometimes they even play the role of agent to connect buyers and sellers.
2. They try to pressure the generation side to lower prices to benefit the local economy, a model that is likely self-defeating if generators are forced to sell at a loss. Instead, generation companies can choose to trade with the grid companies at regulated tariffs if they do not want to trade in the exchange centres or sell directly to large consumers.

Market-oriented pricing mechanisms are the real motivation for generators to enter the market because they have the freedom to explore more profitable opportunities. If the electricity prices in the deregulated market are continuously lower than regulated tariffs, generators will lose interest in market trading. Even though this round of reform has emphasised the importance of markets, administrative influences are still common to see.

The opened-up retail market has attracted a large number of participants within a short period of time, yet the retail sector as a whole has not taken off as expected. The rules and regulations of the retail market need further clarification to provide clearer guidance to participants and to avoid rent-seeking activities.

Currently, many disputes concentrate on settlement mechanisms. The supporting policy document assigned grid companies the responsibility for settlement services, including collecting fees, settling, clearing cross-subsidies, collecting governmental funds (as agent), and paying generation companies and retail companies. Even for generators and consumers/retailers that trade directly with each other, the grid companies are still needed to complete the trade. Grid companies not only enjoy strong liquidity but also hold power over other participants. According to a Regulation Notice released by the NEA in March 2017, some grid companies are in default over payments to generation companies for as long as a year. These late-payments have accumulated to about RMB 1 billion. To make things more complicated, while some companies are not being paid on time, other generation companies are being paid before the power exchange centres have issued settlement documents [25]. These issues might be addressed and prevented with stricter regulations in the future. However, a more practical strategy for generation companies at the moment is to maintain good relationships with grid companies in order to maintain cash flows.

Retail companies that are independent of the grid companies are not qualified to issue receipts to the generation companies with which they sign contracts and from which they purchase electricity. Only the grid companies have the right to collect and pay fees or issue receipts, despite not being party to the trade or contract. This type of participation seems both unnecessary and unnatural under normal market mechanisms: it can be considered a large compromise in this round of electricity market reform. However, compromises are inevitable in macro reform programmes involving various stakeholders nationwide, and in particular, when many of these stakeholders play critical roles in the nation's economy and society.

Currently, retail companies mainly compete on price and gain profits from price gaps between buying and selling. This single business model has worked well for some companies, but the market is still in its initial phase and is yet to be fully established. It is unlikely to be a sustainable model in the long run after market rules are improved and extra profits are fully exploited. This is especially true for independent retail companies that have few advantages when competing with grid-affiliated retail companies on price. Instead, they may benefit from greater focus on providing services supported by innovative technologies and advanced management.

As more problems and issues present themselves to policymakers, the need to clarify the regulations on rights and responsibilities in a more systematic way, rather than solving the issues case by case, will grow. The NEA has listed several measures to prevent unfair competition between grid-owned retail companies and other retailers, including the setting up of power exchange centres with mixed ownership [26]. Currently, as of late 2017, it is too early to examine the effectiveness of these measures. The regulatory framework needs to balance the interests of different stakeholders in the market; otherwise, non-state-owned companies will leave the market within a short period, leaving an 'opened up' market with only state-owned actors. The participation of private and foreign investors in a market can help ensure that companies operate and interact with each other transparently, improving the efficiency of resource allocation in the market. It is critical to build an open and fair regulatory framework and set of market rules, so that private and foreign investors can enter, stay and thrive.

CHANGES TO MARKET RULES

Reform of Transmission and Distribution Tariffs

Historically, electricity tariffs have been fully regulated in China, and tariffs in the T&D sector have never been clarified: grid companies (the T&D service providers) earn their profits by subtracting the on-grid tariff they pay the power plants from the sales price they receive from consumers/retailers. As each power grid company is responsible for a different region, they are the monopolist in both upstream and downstream markets.

The current round of reform has stated that only transmission and distribution, which are natural monopolies, should remain regulated. Nonetheless, the business model of grid companies will need to change from that of electricity buyer and seller to T&D service provider. In future, they will only be able to charge for the T&D service they provide and this will be at regulated T&D tariffs. Power generation companies can trade directly with qualified buyers through the power exchange centres. However, the grid companies' role as the only electricity buyer and seller in the market has hardly changed for they hold the settlement rights and own most of the power exchange centres.

In contrast, reforms regarding the auditing, examining and approving of T&D tariffs have achieved more progress. Pilot reforms started in Shenzhen in January 2015; by June 2017, the reforms had spread to all the grid companies at the provincial level (except for the Tibetan power grid). The T&D tariff examination has saved a large amount of money for consumers by clearing away unnecessary charges. More importantly, costs in the T&D sector have been a black box for decades; this round of audit and approval can finally shed some light on cost structures, improving the transparency of the T&D sector for policymakers and market participants. Given transparent T&D tariffs, which are fully regulated by the NDRC, generators and large consumers/retailers will be able to discover the price in the deregulated market, providing a basis for building up the market-oriented pricing mechanism in the future.

However, complex cross-subsidies still occur within the electricity tariff system, and the solutions for removing them remain unclear. Currently, grid companies are the agent for the collection and payment of these subsidies. In order to set up a market-oriented pricing mechanism, a major reform of subsidies in the power sector will need to be carefully designed and implemented.

Electricity Generation and Consumption Plan

In China, generation rights (for operating hours of generation capacities) have been allocated based on fully regulated electricity dispatch plans for each region and power plant. These dispatch plans are first formed by grid companies based on demand forecasts and then approved by the regulator before being carried out. The number of operating hours a generation company receives determines its profits. Because the grid companies are responsible for planning and executing dispatch plans, they hold great sway over the profitability of generators. Negotiations between the regulator, grid companies and generation companies are meant to help balance the interests of all parties. Because the largest five generation companies in China are all state-owned, they have similar bargaining powers to each other. The annual operating hours are often evenly allocated among these large state-owned generation companies. However, the grid companies responsible for preparing the plan certainly have higher market power than the generators, especially the smaller ones.

The supporting policy document, *Implementation Opinions on the Sequential Opening Up of Generation and Consumption Plan*, issued in November 2015, clearly stated the goal of setting up market-oriented mechanisms to balance electricity supply and demand, and move away from this planned approach. In March 2017, the NDRC and NEA jointly released another document, *Notice of Sequentially Opening the Power Generation and Consumption Plan* [27] that confirmed policymakers' determination in proceeding in this field, and clarified three key rules with regard to the future reform of dispatch plans:

1. Coal-fired power plants are encouraged to trade directly with large consumers or retailers, especially in the form of mid- and long-term contracts.
2. New coal-fired power plants (built after the release of *Document No. 9*) are not regulated by dispatch plan or current electricity tariffs, and should enter the market and trade with buyers directly. Those that have already entered the deregulated market cannot return to the regulated market.
3. Policymakers will sequentially develop and implement the reform of cross-province and cross-region power dispatch plans.

Meanwhile, a large group of consumers remain in the regulated market, including agricultural, residential and some public sectors, which together account for about 30 per cent of China's total electricity demand [18].

CHALLENGES AND OPPORTUNITIES FOR FUTURE REFORM

Compared to the electricity market reforms implemented in 2002, the round of reform that started in 2015 has made remarkable progress in introducing new participants and in changing the rules of electricity pricing and trading. Importantly, major policies and guidance are documented in regulation and legislation at both central and local government levels. This strong policy support has helped the reforms move forward. From *Document No. 9* to its six supporting documents, and many other publicly released notices, policymakers have clearly shown their determination in promoting these new reforms.

Two key factors for examining whether a sector is deregulated are its pricing mechanism and entry qualifications. If the government controls prices and approves/disapproves the entry of market participants in a sector, it is under full regulation. If not, it is deregulated (or partially deregulated). In the current round of reforms, policymakers are trying to set up a transparent T&D tariff system and encourage market participants to decide electricity prices (both wholesale and retail) in the deregulated market on their own. New retail companies need to register with the relevant government agencies and follow certain procedures to enter the retail market, but they do not need to gain administrative approval from any agencies. This round of reform has focused on the two key components of the marketisation of the electrical power industry. However, it faces more challenges in the future.

The principal guideline of this reform is 'control and regulate the middle, open and deregulate both ends', where middle refers to the T&D sector and the ends refer to generation and the wholesale and retail markets. Theoretically, grid companies should only provide T&D services, charging fees at regulated levels. However, they are not only allowed to participate in power selling, but can also own the system operators and most of the power exchange centres. Further, grid companies are the only organisations qualified to provide settlement services and issue receipts, even when they do not participate in the power trades or contracts. In conclusion, the grid companies still cover all the business segments of the electrical power sector except for generation, and hold influence over other participants through their roles as system operator and power exchange and settlement agent. For example, although generation companies can sell electricity to independent retailers, they can only get paid by the grid companies. This affects private and foreign investors in retail markets since they have to compete with retail companies affiliated with grid companies.

The biggest challenge policymakers face may be how to balance market power among different stakeholders, while ensuring that the reforms still proceed smoothly. During the reform process, electricity supply security needs to be ensured and price fluctuations need to be smoothed to support stable economic growth and social stability. Benefits and rights must be shared fairly among the various parties to ensure their participation in the market as well as their support for the reforms overall. This is particularly important for private and foreign investors, as they are more sensitive to costs and profits, and are likely to have greater market and financial discipline than SOEs. Meanwhile, grid companies still hold substantial technical, market and political power, enabling them to wield significant influence over the policymaking process. As a result, grid companies retain greater market power than other sector participants. So far, it is not clear that how the imbalance in market power among different stakeholders can be corrected, and it will be a key challenge for future reform.

References

1. NDRC (2008) http://en.ndrc.gov.cn/mfod/200812/t20081218_252212. html, date accessed 20 July 2017.
2. State Grid Corporation of China (2017) http://www.sgcc.com.cn (home page); China Southern Power Grid, http://eng.csg.cn (home page), date accessed 20 July 2017.
3. Inner Mongolia Power (Group) Co., Ltd, http://www.impc.com.cn/lmsy/ index.htm, date accessed 20 July 2017.
4. The Central People's Government of the People's Republic of China (2011) http://www.gov.cn/zwgk/2011-06/29/content_1895729.htm, date accessed 20 July 2017.
5. China Electricity Council (2017) http://www.cec.org.cn/yaowen-kuaidi/2017-01-25/164285.html, date accessed 20 July 2017.
6. NDRC (2013) http://www.gov.cn/zwgk/2013-07/08/content_2442397. htm, date accessed 20 July 2017.
7. NDRC (2014) http://jgs.ndrc.gov.cn/zcfg/201401/t20140128_577701. html, date accessed 20 July 2017.
8. NDRC (2009) http://www.ndrc.gov.cn/fzgggz/jggl/zcfg/200907/ t20090727_748211.html, date accessed 20 July 2017.
9. NDRC (2016) http://www.ndrc.gov.cn/zwfwzx/zfdj/jggg/201612/ t20161228_833062.html, date accessed 20 July 2017.
10. NDRC (2013) http://www.ndrc.gov.cn/zwfwzx/zfdj/jggg/201308/ t20130830_556127.html, date accessed 20 July 2017.
11. NDRC (2010) http://www.nea.gov.cn/2010-07/28/c_131097727.htm, date accessed 20 July 2017.
12. NDRC (2013) http://www.gov.cn/zwgk/2013-06/09/content_2423501. htm, date accessed 20 July 2017.

13. China Electricity Council (2017) http://www.cec.org.cn/guihuayutongji/gongxufenxi/dianliyunxingjiankuang/2017-07-19/171012.html,   data accessed 10 October 2017.
14. National Energy Administration (2011) http://www.nea.gov.cn/2011-08/17/c_131054463.htm, data accessed 10 October 2017.
15. The Central People's Government of the People's Republic of China (2013) http://www.gov.cn/jrzg/2012-12/23/content_2296789.htm,   data accessed 10 October 2017.
16. National Development and Reform Commission (2004) http://www.ndrc.gov.cn/rdzt/hjpj/200506/t20050615_7664.html, data accessed 10 October 2017.
17. NDRC (2015) http://jgs.ndrc.gov.cn/zcfg/201512/t20151231_770446.html, data accessed 10 October 2017.
18. Liu Xiying & Kong Lingcheng 'A New Chapter in China's Electricity Market Reform', *ESI Policy Brief*. http://esi.nus.edu.sg/docs/default-source/esi-policy-briefs/a-new-chapter-in-china-s-electricity-market-reform.pdf,   data accessed 23 October 2017, 21 March 2016.
19. China Energy Statistical Yearbook 2015 (China Statistics Press), ASIN: B01B5H3WFQ, June 2016.
20. China Electricity Council (2015, 2016) http://www.cec.org.cn/guihuayutongji/gongxufenxi/dianliyunxingjiankuang/2015-02-02/133565.html, http://www.cec.org.cn/yaowenkuaidi/2016-02-03/148763.html,   date accessed 20 July 2017.
21. NDRC (2016) http://tgs.ndrc.gov.cn/zywj/201601/t20160129_773852.html, date accessed 20 July 2017.
22. NDRC   (2015)   http://www.ndrc.gov.cn/zcfb/zcfbtz/201511/t20151130_760016.html, date accessed 20 July 2017.
23. China   Power   (2016)   http://www.chinapower.com.cn/cio-data/20161215/71943.html, date accessed 11 November 2017.
24. Beijixing   Shoudian   (2016)   http://shoudian.bjx.com.cn/news/20170531/828193.shtml, date accessed 11 November 2017.
25. NEA (2017) http://zfxxgk.nea.gov.cn/auto92/201703/t20170330_2756.htm?keywords=, date accessed 20 July 2017.
26. NEA (2015) http://www.nea.gov.cn/2015-12/01/c_134872239.htm, date accessed 20 June 2017.
27. NDRC   (2017)   http://www.ndrc.gov.cn/gzdt/201704/t20170410_843778.html, date accessed 20 June 2017.

CHAPTER 4

Wind and Solar Power in China

Sufang Zhang

Abstract China's renewable energy has witnessed rapid growth since the implementation of its *Renewable Energy Law* in 2006. By the end of 2016, China had the world's largest wind (169 GW) and solar (77.7 GW) capacities along with highly competitive wind and solar PV manufacturing. A policy framework for renewable energy provides targets, mandatory connection and full purchase, categorized on-grid tariffs, cost-sharing mechanisms and a special development fund. Curtailment persists; the main causes lie in transmission bottlenecks, an inflexible generation mix and institutional constraints (such as minimum generation quotas, regulated pricing, within-province supply-demand balancing, and poor compensation for ancillary services). Ongoing market reforms and renewable energy policies aim to address these constraints; success will depend on effective implementation.

Keywords Wind power • Solar power • Renewable energy curtailment • Electricity market reforms • China

S. Zhang (✉)
School of Economics and Management, North China Electric Power University, Beijing, China

© The Author(s) 2018   83
L. Lester, M. Thomas (eds.), *China's Electricity Sector*,
https://doi.org/10.1007/978-981-10-8192-7_4

Summary

China's renewable energy capacity, especially that of wind and solar, has witnessed rapid growth since the implementation of its *Renewable Energy Law* on 1 January 2006. By the end of 2016, the total installed capacity of wind and solar power in the country had reached 169 GW and 78 GW respectively, in both cases the largest of any country in the world. Their combined share of China's total electricity generation in 2016 was 5 per cent (4 per cent for wind power, 1 per cent for solar power). Meanwhile, the competitiveness of wind and solar PV manufacturing industries has also improved dramatically. Chinese wind turbine and solar PV manufacturers are currently taking a dominant position in world markets.

These achievements result from the large number of China's incentive policies, particularly the *Renewable Energy Law*. The *Renewable Energy Law* provides a policy framework consisting of five main elements. These include development targets for renewable energy that provide project developers with clear signals of the government's determination to promote renewable energy; the mandatory connection and full purchase of renewable energy; and categorised on-grid tariffs for renewable energy. These ensure renewable energy project developers can sell all output at fixed prices, guaranteeing investments a known rate of return. There are also cost-sharing mechanisms and a special fund to provide financial support for renewable energy projects.

As the penetration of renewable energy in the power system has grown, so has the curtailment of wind and solar energy, presenting a significant challenge for China's low-carbon ambitions. There exist various causes for curtailment: transmission bottlenecks; an inflexible generation mix; and, most importantly, institutional constraints that include the minimum generation quota allocation system (which gives coal-fired power units priority dispatch over renewable generation), regulated electricity pricing mechanisms (which perversely disincentivise coal-fired power plants from operating more flexibly), within province 'balancing' (which limits interprovincial electricity trading) and the lack of formal compensation mechanisms for electricity ancillary services (which discourages the provision of electricity ancillary services).

In March 2015, China released *Document Number 9*, launching a new round of electricity market reforms characterised as 'deregulation of electricity generation and retail segments and regulation of transmission and distribution'. One of the key tasks of the new reforms was to promote renewable energy development and, in particular, to address the existing institutional constraints. The reform documents, and the new renewable energy policies put in place subsequent to the launch of the new reforms, are promoting China's renewable energy integration through market approaches. These include the application of markets, the full purchase of renewable energy generation according to government issued quotas, the improvement of electricity ancillary services markets, the establishment of a green electricity certificate system and the orderly deregulation of planned electricity generation and consumption quotas. Theoretically, these will help address the challenge of renewable energy curtailment; the actual impact, however, depends on how well the new reforms, and the new renewable energy policies, are implemented.

KEY INSIGHTS

- China's wind and solar power have witnessed dramatic growth since 2006 under the influence of various incentive schemes, reaching almost 250 GW by the end of 2016, making China the world leader in wind and solar installations. At the same time, the country's wind and solar manufacturing industries have also expanded rapidly to become global leaders, helping to drive costs down.

- The *Renewable Energy Law* provides the fundamental policy framework to attract investment and facilitate growth. In addition to capacity targets designed to drive continued expansion, steps were taken to mandate grid connection for renewable assets and prioritise the purchase of renewable generation. Pricing and cost-sharing mechanisms were also overhauled to further encourage development. The on-going new round of electricity market reform is designed to move the power sector governance structure from centralised planning to market-based instruments.

- Wind and solar power curtailment pose a big challenge for the industry, wasting both energy and money, and undermining efforts to reduce reliance on traditional thermal power plants. A variety of interacting forces are behind China's persistent curtailment, including transmission bottlenecks, an inflexible generation mix and a set of institutional constraints that disincentivise the prioritised use of renewable energy. The market forces being introduced through the current reform programme are expected to help ease curtailment.

INTRODUCTION

Energy is vital for economic growth, national security and sustainable development. However, the exploitation of conventional fossil fuels such as coal, oil and gas has led to many environmental problems. Social and political attitudes are forcing environmental and sustainability concerns up the priority list, meaning that economics is no long either the sole or, even, the most important concern. Wind and solar power have become widely used in the electricity sector, and China, the world's largest emitter of greenhouse gases and consumer of coal, has made tremendous effort in developing and installing them. The country is now the world's largest wind and solar power market.

Since the implementation of China's *Renewable Energy Law* on 1 January 2006 [1], a series of government policies have incentivised renewable energy development throughout the country. Consequently, China's wind and solar power have witnessed dramatic growth: its on-grid installed wind capacity has been the world's largest since 2010 [2]. However, along with this dramatic growth in wind and solar generation, curtailment has posed an increasingly serious problem in the country. In this paper, renewable energy curtailment refers to a mandatory reduction in the output of a renewable generator from what it could otherwise produce given available resources.

In March 2015, China launched a new round of power sector market reforms [3]. One of the drivers of this new reform agenda was the increasing penetration of renewable energy. The reform agenda has been characterised as 'deregulate the two ends, regulate the middle', meaning that competition should be introduced into the generation and retail segments, while the transmission and distribution segment remains regulated. Following the overall initiation of the new reform agenda, policies and regulations relevant to renewable energy have been implemented over the past two years.

This chapter looks into the evolution, policies, challenges and opportunities for wind and solar power in China. To this end, the remainder of this chapter is organised as follows: the next section covers the evolution of China's wind and solar PV power sectors. The chapter then sets out the policy framework for renewable energy, including development targets, the full purchase of renewable output, categorised on-grid tariffs (with a focus on feed-in tariffs), cost-sharing mechanisms and the special fund for renewable energy. Renewable energy curtailment and the causes for it are then discussed, focusing on institutional causes including the minimum generation quota allocation system, the regulated electricity pricing mechanism, within-province 'balancing' and the lack of formal compensation mechanisms for electricity ancillary services. Given that one of the key reform objectives of China's on-going power sector reforms is to promote renewable energy development, the chapter then analyses the opportunities provided by the reforms, documenting the main elements relevant to renewable energy integration and other new renewable energy policies.

THE EVOLUTION OF CHINA'S WIND AND SOLAR PV POWER SECTORS

The Evolution of China's Wind Power Sector

China is very rich in wind resources. Its total wind resources are close to the USA's. The evaluation report released by the China Meteorological Administration in 2014 showed that China's onshore technically exploitable wind resources with a wind power density of over 150 W/m^2 at a height of 70 meters reached 7.2 TW [4]. One of the characteristics of wind resources in China is the geographical mismatch between where the wind blows and where the demand for electricity is. Whilst coastal areas in China have huge power demand loads, they have only poor wind resources; in contrast, the 'Three Norths' Region (containing Northeast China— Heilongjiang, Jilin and Liaoning; North China—Inner Mongolia, Hebei and Shanxi; and Northwest China—Xinjiang, Gansu, Ningxia and Shaanxi) has plentiful wind resources but a relatively small power load. This poses difficulties for the economic development of wind power: long distance transmission must be factored into wind project economics.

China has been developing its wind resources for over two decades, but by the end of 2005 had still only 1.3 GW of installed capacity [5]. Since the implementation of the *Renewable Energy Law*, on 1 January 2006, China has made rapid strides in harnessing its wind resources, doubling its installed capacity of wind power each year for four consecutive years [5]. By the end of 2016, total installed wind capacity in the country had reached 169 GW, of which 149 GW had been integrated into the grid (Fig. 4.1) [5]. In consequence, wind generation has grown in both absolute (TWh) and relative (percentage) terms. As a share of China's electricity generation, wind has grown steadily from 0.1 per cent in 2006 to 4.02 per cent in 2016 [6].

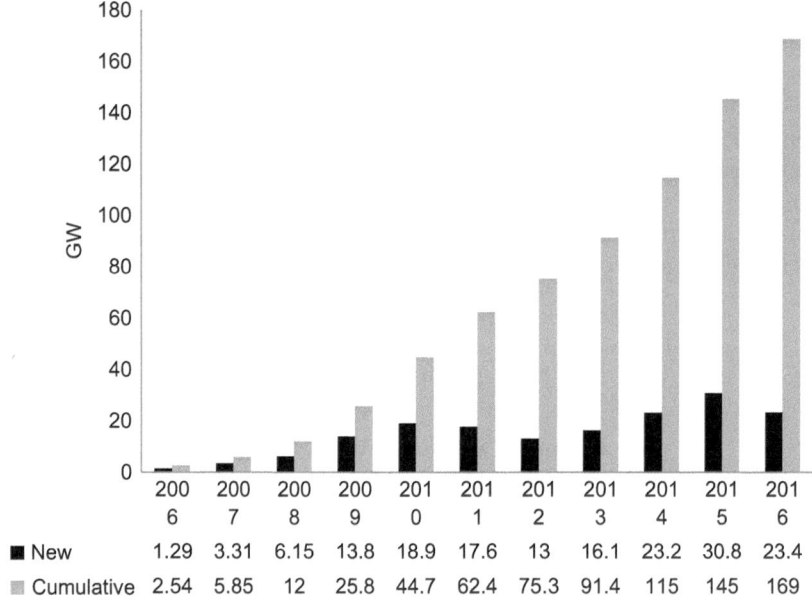

Fig. 4.1 Installed capacity of wind power in China: 2006–16 (Source: China Wind Energy Association [5])

Alongside the scaling up of China's wind power sector, the country's wind turbine manufacturing industry has been developing strongly. In 2016, of the world's top ten wind turbine manufacturers, three were Chinese, with a combined global market share of 20 per cent [7]. Thanks to the Chinese government's support for technological innovation, China's domestic wind turbine technology has improved markedly, and the capacity of individual turbines has grown.

The Evolution of China's Solar PV Power Sector

China also has abundant solar resources: annual sunshine hours and radiation in more than two-thirds of its territory are over 2000 hours at 5000 MJ/m^2 [8]. Technical potentials of utility and rooftop solar PV power amount to 2200 GW and 500 GW respectively [8]. However, the best solar resources in China are largely located in China's north and northwest regions (Xinjiang, Ningxia, Shaanxi, Gansu), far away from load centres [8].

Compared to other leading solar PV countries, such as Germany, Spain, the USA and Japan, China is a relative latecomer to the solar PV industry, particularly in terms of power generation. Nevertheless, the solar PV industry in the country has witnessed dramatic growth since the 2000s [9]. During 2003–08, huge demand from European countries such as Germany drove the expansion of China's domestic solar PV manufacturing industry; in contrast, low domestic demand limited the growth of installed capacity [9]. Before 2009, China's solar PV industry had been export-oriented, with around 95 per cent of Chinese solar modules exported to overseas markets [9].

Since 2009, when overseas demand dropped dramatically because of the 2008 world financial crisis and European anti-dumping and other countervailing measures were placed on Chinese solar PV products, China has put in place incentives to promote its own domestic solar PV market. Since then, annual and cumulative installations of solar PV power in the country have seen unprecedented growth. By the end of 2016, China's cumulative installed solar PV capacity amounted to 77.7 GW, the largest in the world [10] (Fig. 4.2).

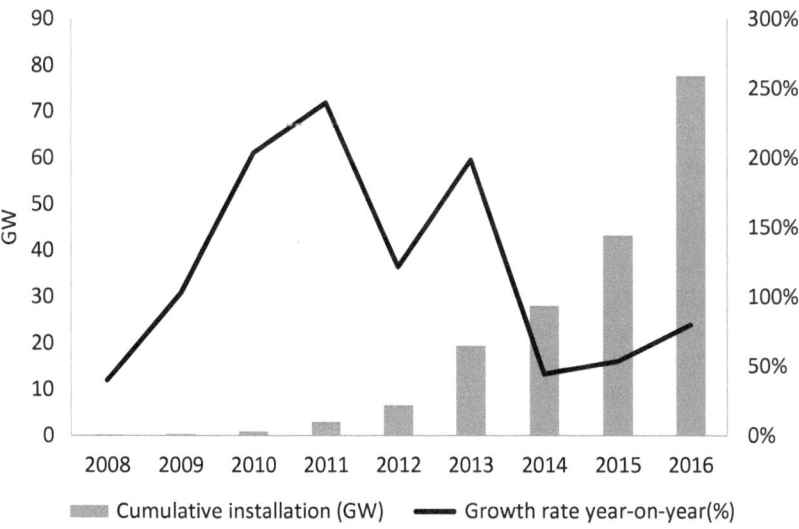

Fig. 4.2 Cumulative installation and growth rate year-on-year of China's solar PV power: 2006–16 (Source: Authors' compilation based on the data released by the NEA)

China's solar PV products currently dominate the global PV industry value chain. In 2016, Chinese output of polycrystalline silicon, silicon wafer, PV cells and PV modules equated to 194,000 MT, 63 GW, 49 GW and 53 GW respectively, accounting for 52 per cent, 94 per cent, 71 per cent and 73 per cent of world production [11]. It is not only in scale that China's PV manufacturing industry has developed; manufacturing technology and efficiency have also been advancing rapidly. The average conversion efficiency rates of mono- and polysilicon cells have increased from 17.5 per cent and 16.5 per cent in 2010 to 19.8 per cent and 18.5 per cent respectively in 2016 [11], taking a leading position in the world. Meanwhile, the costs of mono-silicon and polysilicon materials, PV cells and modules have all witnessed a remarkable decrease. The price of silicon-made modules fell from RMB 12/W in 2010 to RMB 3.2/W in 2016: a 73 per cent decrease [11]. Consequently, the cost of solar PV systems dropped to RMB 7.3/W in 2016. The overall generation cost for solar PV power in China fell by over 60 per cent during the 12th Five-Year Plan (2011–15) [11].

Policies for Renewable Energy Development in China

China's impressive development of wind and solar power since 2006 is attributable to a series of government incentives. The most important of these is the *Renewable Energy Law*, which took effect on 1 January 2006, and was amended in December 2009. The law is significant in that it provides legal support for renewable energy development. It established the following five important policy frameworks for China's renewable energy development:

1. development targets for renewable energy;
2. mandatory connection and full purchase of renewable energy;
3. categorised on-grid tariffs for renewable energy projects;
4. cost-sharing mechanisms; and
5. a special fund for renewable energy development [1].

Details of these five policy frameworks are provided in the following subsections. The law is an umbrella piece of legislation. Details concerning implementation are contained in supporting regulations and measures issued by relevant government agencies such as the National Development and Reform Commission (NDRC) and the National Energy Administration (NEA).

Development Targets

Development targets for renewable energy are intended to give the market a clear signal of the scale of government ambition, unlock investment and help achieve the necessary scale. The targets for renewable energy have been set in China's various renewable energy development plans including the Medium- and Long-Term Development Plan, and the Five-Year Development Plans for renewable energy as well as for each individual renewable energy technology. Over the past decade, it is evident that all the development targets set by these plans for wind and solar power have been reached ahead of time.

In the relevant plans, the share of renewable energy in China's total energy consumption was targeted to reach 10 per cent by 2010 and 15 per cent by 2020. Further, it was required that the two main grid companies were to ensure that 1 per cent of total output within their region came from non-hydro renewable power resources by 2010 and 3 per cent by 2020. Generators with the authority to install and run assets over 5 GW in size were required to achieve 3 per cent of their total installed capacity from non-hydro renewable energy by 2010, and 8 per cent by 2020 [12].

Mandatory Connection and the Full Purchase of Renewables

Mandatory connection and the full purchase of renewable output rules are intended to enhance the financial performance of renewable energy projects, helping to attract project investment. The *Renewable Energy Law* and other relevant regulations and measures specify that grid companies are obliged to connect and purchase the entire amount of electricity generated from renewable energy projects [1]. In practice, however, grid companies often do not fulfil these obligations, arguing that doing so could possibly destabilise the grid. The 2009 amendments to the *Renewable Energy Law*, which took effect on 1 April 2010, seek to improve implementation by balancing the grid company's obligation to connect and fully purchase renewable power with the renewable power generators' responsibility to meet technical standards for on-grid electricity [13]. Instead of simply requiring grid companies to purchase without condition all available renewable power, the 2009 amendments limit the grid company's responsibility to purchase renewable power only from those renewable power generators that meet certain technical requirements for connection [13].

Categorised On-grid Tariffs for Renewables

According to the *Renewable Energy Law*, on-grid tariffs for renewable energy projects are categorised and set by the relevant government agency under the State Council in accordance with the different characteristics and geographies of renewable energy projects. In the early years following the enactment of the law, the on-grid tariff for wind power was determined by the government based on the results of national concession tenders [1]. Building on this tariff-setting practice, in August 2009 the NDRC announced a four-category feed-in tariff (FIT) system for new onshore wind power projects [14]. In determining the FIT in each category, the factors taken into account included local wind resources, the investment cost and an 8 per cent internal rate of return (IRR) [15].

Under the FIT system, the generators of onshore wind power would be paid at a fixed on-grid tariff of 0.51, 0.54, 0.58 and 0.61 RMB/kWh for wind power projects listed as Category I, II, III and IV respectively [14]. Each category covers a different set of cities across the country. Category I covers cities in Xinjiang and Inner Mongolia in Northwest China; Category II covers cities in Hebei, Gansu and Inner Mongolia in North and Northwest China; and Category III covers cities in Jilin, Heilongjiang, Gansu, Xinjiang and Ningxia in Northeast and Northwest China; all other cities are Category IV. Along with the scaling-up of wind power and its subsequent declining generation cost, in December 2015 the NDRC lowered the four categories of FIT to 0.47, 0.50, 0.54 and 0.60 RMB/kWh for onshore wind power projects approved after 1 January 2016 and 0.44, 0.47, 0.51 and 0.58 RMB/kWh for onshore wind power projects approved after 1 January 2018 [16]. Similarly, in August 2013, the NDRC announced a three-category FIT system for ground-mounted solar PV stations, set at 0.90, 0.95 and 1.00 RMB/kWh [17]. In December 2015, these were lowered to 0.80, 0.88 and 0.98 RMB/kWh [16].

Cost-Sharing Mechanism

Under the cost-sharing mechanism, the costs of generation and grid connection are shared by utilities and electricity end-users. The costs of renewable energy tend to be higher than those of conventional energy sources when externalities such as environmental costs are not taken into consideration. As the development of renewable energy benefits the whole of society, it is justified that the cost of developing renewable energy is shared by the whole society.

According to the *Renewable Energy Law* and associated relevant regulations and measures, the additional above benchmark-tariff FIT costs for desulphurised coal-fired power, the costs of public operation and maintenance of independent power systems for renewable energy and the grid-connection costs of renewable energy generation projects are all covered by a national electricity surcharge for renewable energy paid by electricity end-users, including wholesale electricity consumers connected to the grid, electricity consumers with captive power plants and large electricity users who purchase electricity directly from generators. The surcharge rate was set initially at RMB 0.001/kWh in 2006 [18] and raised several times, most recently to RMB 0.019/kWh in 2016 [19].

According to the *Provisional Management Measures on the Imposition and Use of Renewable Energy Development Fund* [20], renewable energy surcharges are to be collected monthly, by both provincial and local grid companies on behalf of the Financial Ombudsman Offices (the Commissioners) of the Ministry of Finance in all provinces, autonomous regions and municipalities directly under the central government. All the surcharge revenue is to go directly to the central treasury.

Special Fund for Renewable Energy Development

In addition to the national electricity surcharge for renewable energy development, the government also offers special financial support to renewable energy stakeholders such as manufacturers and research institutions (among others), to support their production and technological innovation in renewable energy technologies.

The scope of the special fund extends to:

1. the demonstration, promotion and construction of key technologies and of the industrialisation of renewable energy and other new energy;
2. the scaling up of the development and utilisation of renewable energy and new energy as well as capacity building the sector through technical training and so on; and
3. other relevant matters approved by the State Council.

To improve the management of this fund and the national electricity surcharge for renewable energy, the fund and the surcharge were combined to make the Renewable Energy Development Fund under the 2009 amendments [1, 20, 21].

The Challenge of Curtailment

The Prevalence of Curtailment

Along with the dramatic growth in the installation of renewable energy in China, serious renewable energy curtailment has been recorded since 2011. During 2011–16, the average wind curtailment rate was 15 per cent nationwide and the total electricity lost was 145.5 TWh, equivalent to about RMB 72.75 billion in financial losses. Average curtailment rates in Gansu, Jilin and Xinjiang amounted to more than 30 per cent in 2015 and 2016 [22]. Figure 4.3 shows the nationwide average curtailment rate of wind power and the total power lost.

Solar PV curtailment has also occurred since 2013; the average rate of solar curtailment nationwide during 2013–16 was 15.5 per cent. Total abandoned solar electricity in 2016 reached 7.042 TWh, equivalent to a financial loss of RMB 5.6 billion. The worst areas for solar curtailment were in Northwest China including Gansu, Xinjiang, Ningxia and Qinghai [23].

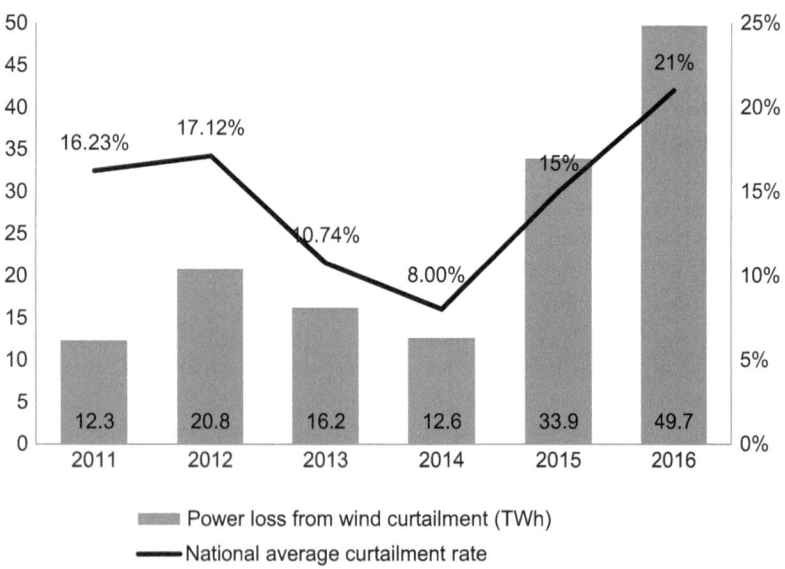

Fig. 4.3 Nationwide average curtailment rate of wind power and power loss incurred (Source: Author's compilation based on the data released by the NEA)

The Main Causes of Curtailment

The increasing curtailment of renewable energy has aroused great concern from government agencies, the renewable energy industry and academics throughout the country. The main causal factors identified include:

- the geographic mismatch between wind and solar resource-rich provinces and China's power load centres;
- the country's inflexible generation mix, which is dominated by coal-fired power;
- the misalignment between transmission capacity and the location of wind farms and large-scale solar stations;
- the lack of energy storage; and
- the recent slow growth of power demand.

Recent research has also suggested that renewable curtailment is rooted in the governance of China's electricity sector, specifically that governance through a planned economy has led to a lack of the flexibility in generation, operation, pricing and demand that is necessary to integrate renewable energy. As such, the key solutions to renewable energy curtailment reside in reforming the existing electricity sector regime. This argument is seemingly increasingly recognised by policymakers. The new round of electricity sector reforms initiated in 2015 states clearly as one of its major objectives the promotion of renewable energy in the country [3]. As such, this section focuses on institutional causes behind renewable energy curtailment.

Transmission Bottlenecks

Transmission bottlenecks are caused by a lack of grid infrastructure to enable long-distance transfer of electricity from China's wind and solar-rich north to its main load centres in the south and east. Wind resources in China are concentrated in the 'Three Norths' Region; solar resources are concentrated in North and Northwest China. According to the *Renewable Energy Law*, the government pays a partial subsidy to the grid companies to link renewable power plants within their service area to the grid as long as they meet technical requirements [1]. In response, the State Grid Corporation of China and the China Southern Power Grid Corporation, China's two main grid companies, have increased their investment in the grid, including the construction and commissioning of 750 kV transmission lines in northwest of China and cross-region transmission capacity expansions in the northwest and northeast [24]. Nevertheless, transmission bottlenecks are still causing curtailment of renewable generation.

Inflexible Generation Mix
China's generation mix is overwhelmingly dominated by coal-fired power, which accounts for 65–70 per cent of total generation [25]. The inherent randomness and fluctuation of intermittent renewable energy sources such as wind and solar require a certain proportion of flexible backup generating capacity in the power system. In China, peaking capacity from flexible power sources such as pumped storage hydropower is less than 3 per cent, in contrast to 34 per cent in Spain and 47 per cent in the USA [25]. Despite low capacity rates among China's thermal power plants resulting from system-wide overcapacity, constraints around ancillary services mean that the available peaking capacity of China's coal-fired power is 50 per cent on average and 20 per cent in heating season, in contrast to 80 per cent in Spain and Denmark [25].

Institutional Constraints
The main institutional constraints to the integration of higher shares of power coming from renewable sources in China lie in the minimum generation quota allocation system, the regulated electricity pricing mechanism, the within province 'balancing' mechanism and the lack of formal compensation mechanisms for electricity ancillary services.

1. *Minimum generation quota allocation system.* Following the unbundling of the grid from generation in the 2002 electricity market reforms, a 'benchmark electricity tariff' for each province was established for thermal generators. This reflected unpublished cost and return expectations, as well as end-user affordability based on the economic development of the provinces. In this context, minimum generation quotas are allocated to coal-fired power units on an annual basis to guarantee sufficient revenue. Such being the case, renewable energy must give way to coal-fired power generation until these minimum quotas are met. In other words, coal-fired power units enjoy real-world priority dispatch, which is inconsistent with the requirement of the *Renewable Energy Law* that renewable energy enjoy priority dispatch [26].

2. *Regulated electricity pricing mechanism.* After the 2002 electricity market reforms, tariffs for generation, transmission and distribution (T&D) and retail continued to be regulated by the central government in China. On-grid generation tariffs for most generation plants were based on 'benchmark' prices designed to constrain end-user electricity prices, while at the same time reflecting local market conditions and promoting (or restricting) particular technologies or fuel types. Electricity retail tariffs were also regulated by the government. The T&D tariffs paid to grid companies were simply the difference between the retail tariff and the on-grid generation tariff [27].

   Coal-fired power generating plants in China are compensated on an RMB/kWh basis (payment is for generation only, there are no capacity payments), and so they strive to generate more power in order to enhance their profits under the minimum generation quota allocation system. This approach to setting tariffs also fails to incentivise coal-fired power plants to operate more flexibly as increasing amounts of renewable energy enter onto the grid.

3. *Within-province 'balancing'.* Large balancing areas with geographic diversity help to reduce the need for flexible reserves. A power market with large balancing areas enables the transmission of power from regions with ample availability of renewable energy to regions with high electricity demand. In China, inter-provincial transmission is commonly fixed for the year through a set allocation mechanism. Regional grids are composed of provincial grids, and most dispatch decisions are made based on balancing production and consumption within these provincial borders. Provinces must craft bilateral contracts (typically annually) stipulating how much electricity can be transmitted across each boundary, except where specific consideration for large energy or transmission projects is given by the central government. This regime is to ensure that all power companies in the province achieve their minimum quota [28] but does little to aid the flow of energy from resource centres to demand centres.

4. *Lack of proper mechanisms for procuring electricity ancillary services.* Electricity ancillary services are a broad range of functions (services) that are required (or procured) by power system operators and provided by network users (generators, customers). These services enable system operators to maintain the integrity and stability of the T&D

system and power quality. In China, ancillary services have long been provided only by grid-connected power plants as either 'basic' ancillary services, which are mandatory and provided free of charge, or 'paid' ancillary services, which receive compensatory payments [29]. Currently, there is no proper mechanism for procuring ancillary services in China and payments to grid-connected power plants for their ancillary services usually account for less than 0.3 per cent of generation revenue, barely covering costs [29]. This has undermined the incentive for power plants to provide ancillary services.

OPPORTUNITIES IN THE CONTEXT OF THE NEW REFORMS

The New Reforms

In March 2015, *Document Number 9* was released by the Central Committee of the Communist Party and the State Council. This signified the launch of a new round of reform for China's electric power sector, following the earlier reforms initiated in 2002. The new reforms have been characterised as 'regulate the middle and deregulate the two ends', meaning progressive introduction of competition into the upstream (generation) and downstream (retail) segments, with continued regulation of the midstream (transmission and distribution). The central element of these reforms was to set transmission and distribution tariffs based on the rule of 'allowable cost plus reasonable profit' [3].

A second key element of the reform was to deregulate the electricity retail business to enable competition, progressively allowing non-state-owned companies to gain access to the electricity retail market, and to enable electricity end-users to choose their retailers. The first step in this was to expand an experiment that was abandoned in 2009 and restarted in 2013 allowing large power consumers to purchase electricity directly from generating companies. Such direct power transactions face many challenges, including the level of transmission and distribution tariffs, the eligibility criteria for generators and the ongoing monopoly position of the grid companies.

Relevance of the New Reforms to China's Renewable Energy Development

To date, the relevance of the new reforms to China's renewable energy development lies in two areas. One is the four initiatives explicitly stated in *Document Number 9* and its six supporting documents for promoting renewable energy development. The other is the new renewable energy policies put in place after the launch of the new reforms.

The four initiatives comprise [3]:

1. prioritised generation for renewable energy entailing a progressive relaxation of the old planned system for generation (which led to fixed generation quotas for conventional fossil fuels), prioritisation of renewable energy in generation planning and dispatch and an increase in the proportion of renewable energy in inter-provincial and inter-regional power trading;
2. the establishment of an ancillary services market to encourage the increased penetration of renewable energy into the power market, to establish a cost-sharing mechanism for ancillary services with power end-users and to proactively carry out inter-provincial and inter-regional trading in ancillary services;
3. regulations on captive power plants to improve the flexibility of the power system and to promote the engagement of such generators in wider ancillary services such as providing peaking capacity, while replacing coal-fired units with renewable energy; and
4. allowing the generators of distributed renewable energy to engage in the retail business.

New Renewable Policies

Pursuant to the new reforms, a series of new renewable energy regulations have been put in place. These include the full purchase of renewable energy generation according to government-issued quotas, priority generation of renewable energy for peaking, reform of the electricity ancillary services market in Northeast China, a green electricity certificate system for renewable energy and the orderly relaxation of the planned electricity generation and consumption quota (Table 4.1).

Full Purchase of Guaranteed Renewable Energy Generation
Document Number 9 made it clear that one of the key undertakings of the new reform is to guarantee the full purchase of renewable energy that meets certain technical requirements for connection. *Document Number 9*'s six supporting documents also provided for priority generation of renewable energy. Nevertheless, specific and operational measures in this respect are lacking. As noted above, in the context of China's

Table 4.1 Major new renewable energy policy pursuant to the new reforms

Document	Policy
Management Measures of the Full Purchase of Guaranteed Renewable Energy Generation (issued on 28 March 2016)	Renewable energy generation consists of guaranteed generation and market traded generation
Trial Measures on Priority Generation of Peaking Units for Renewable Energy (issued on 14 July 2016)	A 'who provides peaking regulation services who benefits' mechanism
Pilot Scheme for Special Reform on Electricity Ancillary Services Market in Northeast China (issued on 18 October 2016)	Establish cost-sharing mechanism for electricity ancillary services in Northeast China
Notification on the Experimental System of Verification, Issuance and Voluntary Subscription and Transaction of Green Power Certificate of Renewable Energy (issued on 6 February 2017)	Voluntary subscription to green certificates
Notification on the Orderly Relaxing of the Planned Electricity Generation and Consumption Quota (issued on 29 March 2017)	Gradually increase market trading power

current power sector regime, it is the conventional thermal plants that enjoy actual priority generation, which has crowded out renewable energy generation and contributed to renewable energy curtailment.

To address this issue, on 28 March 2016, the government promulgated the *Management Measures of the Full Purchase of Guaranteed Renewable Energy Generation* [30]. In these *Measures*, guaranteed generation hours are determined by the NEA and the Economic Operation and Adjustment Bureau under the NDRC, through the principle of ensuring a specified IRR for renewable generators [31]. Shortfalls between actual generation and the generation allowance guaranteed by the government are to be compensated for by generation units at a lower generation merit order, such as coal. Meanwhile, when generation exceeds the guaranteed generation, the excess can be traded in the market. The price of renewable energy generation traded in this way is made up of the competitive market price plus the difference between the local feed-in tariff for renewable energy and the local benchmark tariff for desulfurisation and denitrification.

Improvement of Ancillary Services Mechanism

The new reforms call for the improvement of compensation mechanisms for ancillary services based on the idea of 'Shared Responsibility and Shared Gains'. Electricity consumers and retailers are encouraged to participate in ancillary services by contracting with either generators or grid companies. On 7 June 2016, the NEA announced the opening up of ancillary markets in the 'Three Norths' Region to energy storage, principally pumped storage hydropower [32].

Further, with the approval of the NEA, on 1 January 2017, a pilot electricity ancillary services market was formally initiated in Northeast China. This new ancillary services market is designed to increase the power system's flexibility to absorb intermittent renewables and maintain flexible peaking power [33]. Five kinds of ancillary services are currently being traded in the market, and more ancillary services are expected to be included in the future. Thermal power plants have also been motivated to participate in the competition for managing peak demand [34].

Establishment of a Green Electricity Certificate System

On 6 February 2017, the NDRC, NEA and the Ministry of Finance announced the creation of a green electricity certificate system. The system is currently being run as a trial on a voluntary basis, and only for onshore wind and solar PV power generators (distributed solar PV power generators excluded). Compulsory purchase of green electricity certificates is scheduled to start in early 2018 [35]. This system is to allow wind and solar PV power generators to obtain revenue in a timelier manner than is currently possible, mitigating the pressures on their cash flow resulting from delayed subsidy payments from the government.

Nevertheless, whether this system will help reduce the high curtailment ratios of wind and solar power remains questionable. One issue is the prerequisite of a Renewable Quota System (RQS), a system similar to the Renewable Portfolio Standard (RPS) in the USA and Renewable Obligation (RO) in the UK. The RQS requires relevant electricity market players to generate or purchase a certain proportion of renewable energy. It is both compulsory and market-based in that participants are penalised if they fail to fulfil their obligations but can buy and sell green certificates through the market. Currently, China lacks both the required legal system and the electricity market conditions to implement an RQS. The existing *Renewable Energy Law* fails to provide specific provisions with regard to obligations, rewards and penalties, though it noted the proportion of renewable energy targeted. China's electricity market is still in the process of being developed.

Orderly Deregulation of Planned Electricity Generation and Consumption Quotas

On 29 March 2017, the *Notification on the Orderly Deregulation of the Planned Electricity Generation and Consumption Quota* was issued by the NDRC and NEA. The main element of this document is to gradually increase the trading of electricity in power markets. Accordingly, the current system of planned generation quotas for coal-fired power plants is to be reduced, and inter-regional and inter-provincial power flows (exports and imports) are to be deregulated and promoted. Clean energy generation falling within the scope of state planning is to be prioritised, and generation quotas are to become tradable in power markets [36].

CONCLUDING REMARKS

Since 2006, China has seen a dramatic growth in its installed capacity and generation of both wind and solar power. This growth has been attributed to several incentive policies, particularly the *Renewable Energy Law*, which provides five important policy frameworks for China's renewable energy development, namely:

- development targets for renewable energy;
- mandatory connection and full purchase of renewable energy;
- categorised on-grid tariffs for renewable energy projects;
- cost-sharing mechanisms; and
- a special fund for renewable energy development.

However, along with the increasing penetration of renewable energy into the electricity system, renewable energy curtailment has posed a big challenge for the entire renewable energy industry. The main causes for curtailment are:

- transmission bottlenecks;
- the inflexible generation mix; and, most importantly,
- institutional constraints including the minimum generation quota allocation system, a regulated electricity pricing mechanism, within-province 'balancing' and the lack of formal compensation mechanisms for electricity ancillary services.

The overall objective of the ongoing new reforms is to establish a market-based power sector. Indeed, 'the deregulation of generation and retail and regulation of T&D' in the reforms and the new renewable energy policies pursuant to the reforms address the existing institutional constraints to renewable energy integration through a market approach. Theoretically, these will help address the challenge of renewable energy curtailment in the country. The actual impacts will depend on how well the new reforms and the new renewable energy policies are implemented and enforced.

REFERENCES

1. National People's Congress (2005) *The Renewable Law of People Republic of China*. http://www.gov.cn/gongbao/content/2005/content_63180.htm
2. China Wind Energy Association (2017) *Bulletin of Wind Power Installation Statistics in 2016*. http://news.bjx.com.cn/html/20170217/808890.shtml
3. Central Committee of the Communist Party of China and State Council (2015) *Several Opinions on Deepening Power Sector Reform*. http://www.gov.cn/xinwen/2015-10/15/content_2947548.htm
4. China National Renewable Energy Centre (2015) *Report on the Development of China's Renewable Energy Industry* (China Economic Press).
5. China Wind Energy Association (2017) *Bulletin of Wind Power Installation Statistics in 2016*. http://bit.ly/2xC2ARR
6. China Electricity Council (various years) *Electricity Industry Statistics*. http://bit.ly/2xBV3mi
7. REN21 (2017) *Renewables 2017 Global Status Report*. http://bit.ly/2ghNrlA
8. National Development and Reform Commission (2011) *China's 11th Five-Year Plan for Renewable Energy Development*. http://bit.ly/2gerEv1
9. Zhang S F, Andrews-Speed P & Ji M Y (2014) 'The Erratic Path of the Low-Carbon Transition in China: Evolution of Solar PV Policy', *Energy Policy*, 67, 903–912.
10. Wang B H (2016) *Solar PV Industry in China: Retrospect of 2016 and Prospect of 2017*. http://bit.ly/2w7XFuo
11. China Solar PV Industry Association and China Development Research of Electric Information Industry (2016) *The Development Roadmap of China's Solar PV Industry in 2016* (in Chinese). http://bit.ly/2gfuksh
12. National Development and Reform Commission (2007) *Mid- and Long-Term Plan for Renewable Energy Development* (in Chinese). http://bit.ly/2webQMx
13. National People's Congress (2009) *Renewable Energy Law of the People's Republic of China (Amendments)* (in Chinese). http://bit.ly/2wz3h1I
14. National Development and Reform Commission (2009) *Notice on Improving the On-grid Tariff Policy for Wind Power* (in Chinese). http://bit.ly/2xCnPmH

15. Shi J L (2010) *China's Renewable Energy Pricing Policy* (in Chinese). http://bit.ly/2wJNu0h
16. National Development and Reform Commission (2015) *Notification of Improving On-grid Tariff Policy for Onshore Wind Power* (in Chinese). http://bit.ly/2xmEiw5
17. National Development and Reform Commission (2013) *State Council, 2013. Opinions on Promoting the Healthy Development of Solar PV Industry* (in Chinese). http://bit.ly/2wyNHTu
18. Jiang L P & Wang Q K (2012) How to Address the Deficit of Green Electricity Subsidy (in Chinese), *Energy Review*, Issue 9, 29–30.
19. National Development and Reform Commission (2015) *Notification on Reducing On-grid Tariffs for Coal-Fired Power and Retail Prices for General Industrial and Commercial Power Users* (in Chinese). http://bit.ly/2weB6Tf
20. Ministry of Finance & National Development and Reform Commission (2016) *Notification on Issues Concerning Raising the Levying Standards for the Renewable Energy Development Fund and Other Matters (Caishui [2016] #4)* (in Chinese). http://bit.ly/2w81L5J
21. National Development and Reform Commission, Ministry of Finance and National Energy Administration (2011) *Provisional Management Measures on the Imposition and Use of Renewable Energy Development Fund (Caizong No. 115 [2011]* (in Chinese). http://bit.ly/2wAlT1b
22. National Energy Administration (from 2011 to 2016) *Situation of Wind Power Integration in China* (in Chinese). http://www.nea.gov.cn
23. National Energy Administration (from 2011 to 2016) *Solar PV Generation Statistics.* http://www.nea.gov.cn
24. State Grid Cooperation of China (2006) *Summary of the State Grid Cooperation of China's 750 kV Transmission Demonstration Project* (in Chinese) (China Electricity Press, 1st ed).
25. Shu et al. (2017) Study on Key Factors and Solutions of Renewable Energy Accommodation (in Chinese). *Proceedings of the CSEE Chinese Society for Electrical Engineering*, 37(1) January 5, 1–8, https://doi.org/10.13334/j.0258-8013.pcsee.162555.
26. International Energy Agency (2006) *China's Power Sector Reforms: Where to Next?* http://bit.ly/2w1Blkp
27. Kahrl F & Wang X (2014) *Integrating Renewables into Power Systems in China: A Technical Primer—Power System Operations* (Beijing, China). http://bit.ly/2w8ke28
28. Depuy et al. (2015) *Low-Carbon Power Sector Regulations: Options for China.* http://bit.ly/2xqoPu1
29. Lu R (2016) 'Preliminary Study on China's Electricity Ancillary Services Market' (in Chinese), *Southern Energy Observation.* http://bit.ly/2g3exwH

30. National Development and Reform Commission (2016) *Management Measures of Full Purchase of Guaranteed Renewable Energy Generation* (in Chinese). http://www.nea.gov.cn/2016-03/28/c_135230445.htm

31. Qin H Y (2016) 'Interpretation of Management Measures of Guaranteed Generation', *Journal of Wind Energy* (in Chinese), April, 24–26.

32. National Energy Administration (2016) *Notification of Promoting Energy Storage Participation in Ancillary Services in the 'Three Norths' Region* (in Chinese), June 7. http://bit.ly/2io4rqI

33. National Development and Reform Commission and National Energy Administration (2016) *Pilot Scheme for Special Reform on Electricity Ancillary Services Market in Northeast China* (in Chinese). http://www.chinasmartgrid.com.cn/special/?id=620584

34. Suo W (2017) *Breakthrough in Ancillary Services in the Process of Electricity Reforms, Northeast China Electricity Ancillary Services Market Ensures Heating Supply, Electricity Supply and Wind Curtailment Reduction* (in Chinese), June 2. http://bit.ly/2vYw8eQ

35. National Development and Reform Commission, Ministry of Finance (MOF) and National Energy Administration (2017) *Notification on the Experimental System of Verification, Issuance and Voluntary Purchase and Transaction of Green Electricity Certificate of Renewable Energy* (in Chinese). http://bit.ly/2iBO3mS

36. National Development and Reform Commission and National Energy Administration (2017) *Notification on the Orderly Deregulation of the Planned Electricity Generation and Consumption Quota* (in Chinese). http://bit.ly/2w1TqhY

CHAPTER 5

Environmental Regulation

Lele Zou and Ying-Zi Wang

Abstract The power sector remains a major source of atmospheric pollution within China. In this chapter, the environmental regulation of the
power sector over the last 30 years is summarised, along with the effect on
power generation, demand and the grid. Five environmental regulations
(replacing smaller units with big ones, ultra-low discharge of coal power,
tradable green certificates, carbon markets, and taxation of energy and
carbon) are picked out for special attention in light of their significant
impacts on the sector. The overall effect of regulations is uncertain because
of uncertainties in new energy investment, coal power emission reduction
controls and business management in the power sector.

Keywords Power sector • Environmental regulation • Carbon market
• Carbon tax • Pollution

L. Zou (✉) • Y.-Z. Wang
Institute of Science and Development, Chinese Academy of Sciences,
Beijing, China

© The Author(s) 2018
L. Lester, M. Thomas (eds.), *China's Electricity Sector*,
https://doi.org/10.1007/978-981-10-8192-7_5

SUMMARY

The challenges China faces in terms of environmental protection have become as well-known as its successes in reducing that environmental footprint. Energy and carbon intensities, emissions of sulphur dioxide and nitrogen oxides and the efficiency of coal consumption in the power sector have all seen dramatic improvements over the span of the 11th and 12th Five Year Plans (FYPs). Nonetheless, China's power sector remains largely reliant on coal-fired capacity and is a major source of atmospheric pollution within the country.

Over the last 30 years, China has published nine laws, over 50 administrative regulations and nearly 200 regulatory documents on environmental protection. Key among these are: the *Environmental Protection Law of the People's Republic of China (Amendment)* and the *Law of the People's Republic of China on the Prevention and Control of Atmospheric Pollution (Amendment)*, which set industrial pollution standards; the *Environmental Protection Tax Law*; and the *Outline of the 13th Five-Year Plan for the National Economic and Social Development of the People's Republic of China*, which sets a series of targets for 2020 including reducing energy intensity by 15 per cent, capping coal consumption at under 5 billion tonnes, reducing carbon dioxide intensity by 40 to 45 per cent and limiting emissions of sulphur dioxide and nitrogen oxides below 15.8 million tonnes.

Supporting such high-level strategic policy documents, China has also introduced a set of management systems including limits on new capacity approvals, environmental impact assessments, pollution permits and tradable rights, environmental protection taxes and tradable green certificates. In addition, various subsidies, dispatch prioritisation and minimum generation quotas are all being implemented with the view to incentivise the rapid expansion of renewable generating capacity.

The overall effect of these various initiatives has been to optimise the generation fuel mix, improving the efficiency and environmental footprint of coal use (such as through the ultra-low discharge standards) but also reducing coal generation in favour of renewables. This has also led to a sustained programme to phase out direct fuel use, replacing inefficient coal boilers in industry and residential properties with electricity. Improved technology standards are also helping in demand-side management: all industrial sectors are required to upgrade old and inefficient equipment, while efficiency standards on products such as lighting and air-conditioning are facilitating energy conservation among the general population. New policies are also attempting to promote smart grids, seen as a key platform for energy conservation.

A nationwide carbon market is expected to raise generators' running costs by about 2.5 per cent, and is an important part of the government's attempt to promote low-carbon energy. A carbon tax is also expected after 2020, which it is hoped will reduce emissions, stimulate the adoption of low-carbon technologies and reduce energy demand, though at the possible cost of increased unemployment within the power sector.

China's growing framework of regulations has certainly helped the country enhance its environmental protection. Nonetheless, the overall effect of regulations is not always certain: renewable capacity has increased rapidly, but generation still lags behind. Regulations to replace outdated small coal-fired units with larger and more efficient plants might actually be worsening overcapacity and business management strategies may not yet have fully incorporated the business risks a low-carbon future can hold for traditional power sector participants.

KEY INSIGHTS

- **For the government:** due to the limited degree of marketisation in China's power sector, there isn't a smooth signalling mechanism (such as electricity price) for producers and consumers to interact with each other. Most of the emission reduction costs are felt at the generation end of the traditional power sector, without a consumer response, limiting the efficiency of various environmental policies. In contrast, the cost of renewable energy construction has largely been carried by the government, in the form of subsidies, reducing the enterprises' motive for technological improvements. Reformed energy and power markets may help improve policy efficiency.

- **For the power enterprises:** although the current market is imperfect, the government is implementing a growing number of more market-oriented policies (including a carbon market, carbon and energy taxes and so on). In order to adapt to these new circumstances, more attention should be paid to emission intensity, technological progress, the level of emissions and other environmental factors affecting costs and profits. Properly accounted for environmental costs could enhance asset management.

- **For the whole electric power sector:** more diversified participants, such as third-party energy/emission management organisations, trading agents, finance and investment institutions and other relevant agencies and departments, if fully incorporated into the emerging power market alongside producers and consumers, could help the sector meet the challenges of future policy and structural shifts afoot in the whole industry.

THE POLLUTION BEHIND CHINA'S ELECTRIC POWER SECTOR

In recent years, great progress has been made in terms of environmental protection in China. During the 12th FYP period (2011–15), the energy consumption per unit of gross domestic product (GDP) decreased by 18.4 per cent; the total discharge of major pollutants, such as sulphur dioxide, ammonia nitrate and nitrogen oxides, decreased by 18.0 per cent, 13.0 per cent and 18.6 per cent, respectively; and the carbon dioxide discharge intensity dropped by over 20 per cent [1]. In the electric power sector specifically, the proportion of coal generation capacity equipped with desulphurisation and denitrification units went up to 99 per cent and 92 per cent, respectively. The amount of coal consumed per kWh decreased by 18 grams of standard coal [2]. Standard coal is also called coal equivalent, which has a unified calorific value. The calorific value per kilogram of standard coal is 7000 kcal. Raw coal is produced by the coal mine without washing or processing. Raw coal can be divided into peat, lignite, bituminous coal and anthracite, based on the degree of carbonisation. Across the board, the development and installation of clean electricity generation technology have been accelerated in China: non-fossil fuel generating capacity accounts for 35 per cent of China's total, with the country's annual additions accounting for roughly 40 per cent of global annual non-fossil installations [3].

It remains undeniable that China is still faced with a severe situation in terms of environmental protection. It is the world's largest emitter of sulphur dioxide, nitric oxide and greenhouse gases. In 2016, 75.1 per cent of Chinese cities failed to meet national air quality standards [4]. The pollution problem has contributed to a significant loss in people's health and to the general economy, and has resulted in growing public pressure from home and abroad.

The electric power sector is an important energy consumer and a major source of pollution. In 2015, total coal consumption in China was 3.97 billion tonnes of standard coal; 1.65 billion tonnes were consumed in the production and supply of electricity and heat: about 42 per cent of the total. Again in 2015, the power sector discharged 5.28 million tonnes of sulphur dioxide and 5.52 million tonnes of nitrogen dioxide, accounting for 28.4 per cent and 29.8 per cent of China's total [5].

Against this backdrop, the Chinese government has committed to pay significantly more attention to environmental protection and to develop a more comprehensive framework of regulation for the power sector. Environmental protection encompasses much more than just atmospheric pollution: China is suffering from water and land pollution as well, and the power sector has been implicated in both. Nonetheless, this chapter will focus on atmospheric pollution.

OVERVIEW OF EXISTING ENVIRONMENTAL REGULATIONS

China's electricity sector is governed both by the country's overarching environmental legislation, and by specific regulations targeting electricity generators (supply side) and end-users (demand side).

Over the past 30 years, China has passed or published nine laws, more than 50 administrative regulations and nearly 200 regulatory documents on environmental protection, along with 15 laws governing natural resources [6].

- Industrial pollution discharge standards were set through the *Environmental Protection Law of the People's Republic of China (Amendment)* and the *Law of the People's Republic of China on the Prevention and Control of Atmospheric Pollution (Amendment)* enforced in 2015 and 2016, respectively [7, 8]. The laws also emphasised the importance of adjusting the energy structure by promoting clean energy in order to reduce pollution.
- On 25 December 2016, the Standing Committee of the National People's Congress passed the *Environmental Protection Tax Law*, the first law specifically designed for green development, to start from 1 January 2018 [9]. As of the time of writing, the details of this law were still under discussion, though the range of the tax rate has already been set within the law.
- In 2010, legislation for the *Law of the People's Republic of China on Responses to Climatic Changes* was drafted; it remains in progress. The law would be the first 'in response to climatic changes' in China. As the electric power sector is a major source of greenhouse gases (GHG), the law is expected to have an important effect on both energy utilisation and the power sector's fuel mix.

Through these various laws and regulations, a series of environmental management systems, including environmental impact assessments, regional approvals, pollution discharge permits, controls over the total emissions of major pollutants, standards for the release of individual pollutants, pollution taxes and trading systems for carbon and other pollutants have all been established in China.

At a more macro level, the government has a number of strategic plans giving long-term guidance (with phased objectives) on both environmental protection and energy use. In the 13th FYP (2016–20), energy and the environment have risen in importance once again. The *Outline of the 13th Five-Year Plan for the National Economic and Social Development of the People's Republic of China* calls for a modern energy system, defined as low-carbon, safe and efficient [10]. As part of this, high-level targets and limits on energy consumption and pollution discharge (including that of carbon) have been set. These include the following examples, all of which are to be met by the end of 2020, against a 2015 benchmark where appropriate:

- energy intensity is to fall by 15 per cent;
- coal use is capped at under 5 billion tonnes of standard coal;
- sulphur dioxide emissions are capped at below 15.8 million tonnes;
- nitrogen oxides emissions are capped at below 15.8 million tonnes; and
- carbon dioxide intensity is supposed to be reduced by 40 to 45 per cent.

China can, therefore, be seen to be putting in place an increasingly comprehensive framework of environmental policies, providing a legal basis for their enforcement as it pushes the electric power sector to adjust its energy structure and reduce its discharge of pollutants.

On the basis of these types of document, the environmental regulation of the electric power sector can be divided into two levels:

1. strategic planning, with long- and medium-term objectives; and
2. management systems and their relevant laws and regulations.

Strategic Plans for Environmental Regulation in the Power Sector

Many of these long-term plans, such as the *Plan for Energy Conservation and Discharge Reduction during the Thirteenth Five-Year Plan Period*, the

Plan for Environmental Protection during the Thirteenth Five-Year Plan Period and the *Plan for the Development of the Electric Power Sector during the Thirteenth Five-Year Plan Period (2016–20)*, place either direct or indirect constraints on the electric power sector [11–13].

There are six key documents that are driving the reshaping of China's energy sector generally, and the way in which it uses coal and reduces its emissions more specifically. All of them set clear targets to be achieved by the end of 2020. Table 5.1 summarises some of the main targets currently in force [14–16].

Management Systems and Related Laws and Regulations

An array of process and management systems have been created to help lessen the environmental consequences of the power sector (Table 5.2) [17–24].

The government has also tried to encourage enterprises to speed up adjustments to the structure of the country's electricity generation sector through a variety of special measures [25, 26] such as:

- new-energy subsidies;
- environment-protecting electricity top-up prices;
- electricity generation quotas; and
- prioritisation of renewable energy dispatch to the grid.

These various systems and measures have been given legal footing through the *Electricity Law of the People's Republic of China,* which states that 'in the construction, production, supply and use of electric power, the environment shall be protected according to law by adopting new technologies to reduce the discharge of harmful substances and prevent pollutions and other public hazards [27]. And the government encourages and gives support to electricity generation by use of renewable and clean energy resources.'

China's environmental laws, management systems and strategic plans have become increasingly comprehensive, and the requirements for energy conservation and pollution reduction within the power sector are increasingly stringent. This has happened alongside changes to the industry's structure and a substantial increase in investment in environmental protection within the power sector.

Table 5.1 Indicators and 2020 targets in environmental plans affecting the electric power industry

Category	Indicator	Target/Limit/Criteria	Source
Energy structure adjustment	Installed hydropower capacity	>350 GW	Energy Development Strategy Action Plan (2014–20)
	Installed nuclear power capacity	>58 GW	
	Installed wind power capacity	>200 GW	
	Installed solar photovoltaic capacity	>100 GW	
	Installed biomass-fired generating capacity power	>30 GW	National Plan in Response to Climatic Changes (2014–20)
	Installed non-fossil fuel generating capacity	>770 GW, to be about 39% of capacity and account for 31% of generation output	Plan for the Development of the Electric Power Sector during the Thirteenth Five-Year Plan period (2016–20)
	Installed LNG-fired generating capacity	>110 GW, to be >5% of total capacity	
	Installed coal-fired generating capacity	<1100 GW, to be <55% of total capacity	

Coal power energy conservation and pollution reduction	Average coal consumption for upgraded coal-fired generation units	<310 g/kWh	Coal Power Energy Conservation, Discharge Reduction, Upgrading and Transformation Action Plan (2014–20)
	Coal consumption for units larger than 600 MW (excluding air cooling units)	<300 g/kWh	
	Coal consumption for newly built generating units	<300 g/kWh	
	Shutdown technologically obsolete and polluting thermal power capacity	if unit size >20 GW	Plan for the Development of the Electric Power Sector during the Thirteenth Five-Year Plan period (2016–20)
	Decrease in thermal power units' annual sulphur dioxide discharge	>50%	
	Decrease in thermal power units' annual nitrogen oxides discharge	>50%	
	Carbon dioxide intensity of coal-fired generation units	<865 g/kWh	
	Proportion of thermal power plants achieving national standards for wastewater discharge	100%	
	Nitrogen oxides emissions from coal-fired boilers	<100 mg/m^3	Emission Standard for Air Pollutants for Thermal Power Plants (2011)
	Sulphur dioxide emissions from new coal-fired boilers	<100 mg/m^3	
	Sulphur dioxide emissions from existing coal-fired boilers	<200 mg/m^3	
	Soot from coal-fired boilers	<30 mg/m^3	
	Mercury and related compounds from coal-fired boilers	<0.03 mg/m^3	

Table 5.2 Management systems and their relevant laws and regulations

Management systems	Interpretations	Laws and regulations
Regional limited approvals	The construction of any coal-fired power plant in urban or suburban areas of large and medium-size cities is banned No new coal-fired projects (or expansions to existing capacity) can be allowed until all existing units comply with sulphur dioxide emissions regulations	Notice on Enhancing the Prevention and Control over Sulphur Dioxide Pollutions Caused by Coal-Fired Power Plants (2003)
Environmental impact assessments	Prior investigation, forecasting and evaluation are to be conducted on the potential environmental damage from activities related to the construction and development of projects Preventive measures are to be taken to mitigate possible effects on the environment Monitoring is to be conducted on all projects under construction	The Environmental Impact Assessment Law of the People's Republic of China (2003, revised in 2016)
Pollutant discharge permits	The existing thermal power generators must have permits to discharge pollutants and must have in place systems for monitoring discharges, information disclosure and routine reporting	Implementation Plan for the Permit System for Controlling Pollutants Emission (2016) Notice on Deploying the Management over the Pollutant Discharge Permit from Elevated Sources in Pilot Cities in Beijing, Tianjin and Hebei Province as well as in the Thermal Power Industry and the Papermaking Industry (2017)
Environmental protection tax	Fees or taxes are collected for the discharge of pollutants above allowed limits	Measures for the Administration of the Charging Rates for Pollutant Discharge Fees (2003) Environmental Protection Tax Law (2016)

(*continued*)

Table 5.2 (continued)

Management systems	Interpretations	Laws and regulations
Pollution rights trading/carbon trading	A government cap is imposed on the total amount of pollution allowed; tradable permits are issued to polluting enterprises	Instructional Advice on Implementing Pilot Work on Paid Use and Trading of Emission Permits (2014) Interim Measures for the Administration of Revenue Management for the Sale of Pollution Rights (2015)
Tradable green certificates (TGC)	Green certificates are issued to renewable energy providers based on their generation. Provinces or corporations which cannot meet the required minimum proportion of renewable energy can buy TGC through trading platforms	Instructional Advice on Establishing the Target Guidance System for the Development and Utilization of Renewable Energy (2016) Notice on the Issue of Renewable Energy Power Green Certificates and its Voluntary Subscription Trading System (2017)

EFFECTS OF ENVIRONMENTAL REGULATIONS ON THE ELECTRIC POWER SECTOR

While the regulations discussed so far have had environmental protection as their main concern, their effects on the electric power sector have reached further than simple emission controls might suggest.

Effects of Environmental Regulations on Generation

Increasingly stringent environmental regulations have caused the sector to engage in long-term planning, accelerating the transformation of the power supply and the promotion of cleaner and more efficient coal use.

The Power Supply Structure Is Being Optimised

Atmospheric pollutants discharged by the electricity sector, such as sulphur dioxide and nitrogen oxides, are mainly attributable to coal. The fundamental measure has therefore been to reduce the proportion of coal-fired power and develop non-coal generation (such as natural gas, renewables, nuclear and other energies with near-zero emissions). This trend is reflected not only in the overall national plan for the power industry but also in the selection of development paths for the electric power generating enterprises themselves.

In 2016, the National Development and Reform Commission issued the *Measures for the Administration of the Guaranteed Purchase of Electricity Generated by Renewable Energy*, specifying that grid companies must prioritise renewable energy ahead of thermal generation [28]. In March 2017, it was proposed in the *Notice on Releasing Electric Power Generation Plans and Electric Power Use Plans in an Orderly Manner* that the planned quota for coal-fired generation should be gradually reduced year on year [29].

At the enterprise level, since 2011, the five major electric power generating groups have begun to sell down their stakes in non-performing thermal electric power assets, while focusing on making investments in new energy projects. It can be seen from Table 5.3 that in recent years the proportion of clean energy (mostly wind and photovoltaics) within the portfolios of the five major power generating groups has increased steadily.

Under the dual guidance of environmental policies and industry development trends, investment in clean energy amounted to USD 10.29 billion in 2015: one-third of total global investment in power generation [30]. Rebounding coal prices in 2016 meant that many coal-fired generators ran at a loss. In addition, the on-going power sector reform measures have increased the constraints on coal-fired power, again shifting the emphasis onto cleaner technologies. It has been forecast that during the 13th FYP period, an additional RMB 2.5 trillion will be invested in renewable energy, up nearly 39 per cent on the 12th FYP period [1].

Table 5.3 Changes in the proportions of the five major power generating groups' installed capacity by use of clean energies from 2011 to 2016 (unit: %)

Company	2011	2012	2013	2014	2015	2016
China Huaneng Group (CHG)	19.1	21.0	24.6	27.1	28.8	29.0
China Datang Corporation (CDC)	21.4	22.8	25.9	28.1	30.5	31.8
China Huadian Corporation (CHC)	25.4	25.0	30.6	34.0	37.1	37.3
China Guodian Corporation (CGC)	21.9	22.5	25.0	26.9	29.9	30.3
State Power Investment Corporation (SPIC)	29.8	31.0	34.2	38.5	40.1	42.9

Source: www.cemr.org.cn/

Coal Power Production Is Becoming Cleaner and More Efficient
It is generally accepted that the thermal power industry in China is suffering from severe over-capacity. In 2016, capacity utilisation hours fell lower than at any point in the previous 50 years, trapping many coal-based power enterprises in financial difficulty [31]. Policies have been announced by the government to limit the capacity expansion of coal plants. In the first half of 2017, investment in thermal power was 17.4 per cent lower than for the same period in 2016; investment in coal power decreased by 29 per cent [32].

Although the operating hours for individual coal power units have become constrained, it remains China's dominant source of power and is, of course, cheap. Room remains for this existing capacity to improve its efficiency and reduce emissions. Environmental regulations such as 'Replacing Smaller Units with Big Ones' and 'Ultra-Low Discharge of Coal Power' are forcing existing coal plants to conserve energy and reduce emissions, and are also closing down technologically obsolete facilities. Under the guidance of these policies, coal power enterprises have improved their efficiency, expanded heat and power cogeneration, improved energy conservation and optimised operations, enabling them to reduce carbon emissions.

Effects of Environmental Regulations on Demand and the Grid

The effects of environmental regulations on electricity demand have been felt mainly through reductions stemming from energy conservation and promotions stemming from electricity replacement programmes. Attention to environmental protection has also provided more opportunities for smart grids.

Improved Technology Standards Are Helping to Reduce Consumption
At present, all industrial sectors are required to eliminate inefficient production capacity and speed up the upgrading of equipment. New energy consumption standards for products have forced traditionally large electricity users (such as the petrochemical, papermaking and textile industries) to improve their energy efficiency. Electricity demand is also being reduced thanks to improved energy efficiency standards for buildings, and the retrofitting of various industrial parks to upgrade energy efficiency.

Additionally, since 2011, *Administrative Measures for Power Demand-Side Management* have required electricity-using products to reduce consumption [33]. Demand-side management programmes aim to cover multiple parties including industry and the grid companies, but also energy service companies and residential and commercial end-users. Improved pricing incentives, low energy lighting and energy-efficient air-conditioners all help raise awareness of energy conservation among the general population and, in consequence, extend energy savings.

Electricity Replacement Programmes Are Phasing Out Inefficient Direct Fuel Use

Direct bulk coal and fuel combustion are the two weakest areas for China in terms of atmospheric pollution control. They are also important causal factors for the poor quality of the atmospheric environment in northern China during winter. Compared with bulk coal, electricity has significant advantages in efficiency, convenience and pollution control. China has carried out an electricity replacement programme to phase out the direct burning of coal and other fuels in industry and elsewhere. In the 13th FYP, this electricity replacement programme has been extended to cover residential heating in northern China, industrial and agricultural energy production and electric vehicles. It is estimated that by 2020, with the promotion of the programme, the share of electricity in terminal energy consumption will increase from 25.8 per cent to more than 27 per cent. This equates to an increase in electricity consumption of about 450 TWh [13].

New Policies Are Promoting Smart Grids and Ultra-High Voltage Transmission

Two emerging features of China's power system that are now being indirectly supported are ultra-high voltage (UHV) transmission and smart grids.

UHV transmission, flexible transmission, low-carbon dispatching and distributed generation can all help reduce energy transmission losses. Smart grids would also aid the structural adjustment of the power sector. Qualified smart grids are the only real way to truly support the large-scale implementation of renewable energy on the grid. They can help users manage their electricity demand throughout the day, reducing peak load and stabilising hour-by-hour load variations. The implementation of the two-way interactions with end-users that are enabled by smart grids can not only improve the efficiency and predictability of demand, but also facilitate the entry of new technologies and practices to the power system, such as the large-scale application of electric vehicles. Smart grids are therefore seen as platforms for energy conservation, improved grid efficiency and enhanced renewable integration. All of this makes them of great interest in environmental protection.

Five Important Environmental Policies

In this section, five existing or soon-to-be-enforced environmental regulations are picked out for special attention as they are believed to either have made or will make significant impacts on the electric power sector.

Replacing Smaller Units with Big Ones

Replacing smaller units with big ones refers to the campaign to replace small inefficient coal boilers with modern, high capacity-high efficiency units. New power supply projects are often conditional upon the decommissioning of small and inefficient thermal power generating units. The *Notice of Several Opinions on Accelerating Close-Down of Small Thermal Power Generating Units* issued in 2007 provided the policy basis for this campaign [34].

The policy elicited extensive responses from both industry and local government. During the 11th FYP period, the total capacity of all small thermal power generating units that were closed across China totalled 77 GW, of which roughly 67–70 GW were coal-fired. During the 12th Five-Year Plan period, a further 28 GW of small capacity thermal power units were retired. As a result, the proportion of units rated supercritical or ultra-supercritical increased significantly, and the proportion of units with a capacity of 300 MW and above increased to 78.6 per cent, with 41 per cent of units larger than 600 MW. It is estimated that this programme alone has led to 69 million tons of raw coal, 1.2 million tonnes of sulphur dioxide and 139 million tonnes of carbon dioxide being saved each year [35].

Ultra-Low Discharge Coal Power

Ultra-low discharge coal power refers to the various technologies available to scrub pollutants from coal-fired emissions. These technologies can even bring discharge standards for coal below those of gas, with sulphur dioxide emissions limited to 10 mg/m^3 and nitrogen oxides to 35 mg/m^3, making it possible to use coal in a clean and efficient manner (excepting carbon). The 'ultra-low discharge' programme has therefore helped to maintain the role of coal-fired power while improving environmental outcomes. It is stipulated in the *Coal Power Energy Conservation, Discharge Reduction, Upgrading and Transformation Action Plan (2014–20)*, issued in 2014, that by 2020 all coal-fired plants must meet the 'ultra-low discharge' standards; the *Program for the Full Implementation of Transformation Oriented at Ultra-Low Discharge and Energy Conservation in Coal-Fired Power Plants*, issued in 2015, is designed to drive implementation [36].

During the Twelfth FYP period, 400 GW of coal-fired capacity was upgraded to meet energy conservation targets, of which 160 GW met the 'ultra-low discharge' standards. The 13th FYP period should bring this 160 GW up to nearly 420 GW, containing an apparently lower target than the *Action Plan* mentioned above [37].

The 'ultra-low discharge' programme also obliges coal-fired power generating units to pursue other technological advances, such as precision ammonia injection and upgraded double-loop desulphurisation technology, both of which reduce both pollution and costs.

Tradable Green Certificates

Tradable green certificates (TGCs) are vouchers based on non-hydro renewable energy power generation output. In 2001, the Netherlands took the lead in issuing green certificates, and similar policies have now been implemented in more than 20 countries. In 2006, *Instructional Advice on Establishing the Target Guidance System for the Development and Utilisation of Renewable Energy* was published by the National Energy Administration, marking the establishment of TGCs in China. It was followed by a more specific instruction, the *Notice on the Issue of Renewable Energy Power Green Certificates and Its Voluntary Subscription Trading System*, which was released in 2017 [21].

From the perspective of government, TGCs are a good alternative to renewable energy subsidies. For thermal power enterprises, TGCs can be an economically efficient way to meet the minimum required renewable energy quotas, which were designed to encourage generators to allocate a larger share of investment to renewable energy and reduce the curtailment of wind and photovoltaic power. TGCs can also improve liquidity among renewable generators, helping to boost investment, improve renewable energy technology and reduce costs. Clearly, certified green power can also be an opportunity for brand promotion.

As of 2017, China's TGC system is still in its pilot stage. Further efforts are needed to establish supporting institutional mechanisms, such as a unified measurement and regulatory mechanism.

Carbon Markets

In 2014, the Chinese government committed to peak carbon dioxide emissions by 2030, with best efforts made to peak as soon as possible. The electric power industry is the largest emitter of carbon dioxide in China, and the sector's carbon intensity is far higher than that for developed countries or even the world average. Transformation of the power sector is key for the Chinese economy to realise a low-carbon future. Carbon trading has been identified as a possible mechanism to help with this transformation.

Since the first pilot carbon market started in June 2013 in Shenzhen, other pilots were started in Shanghai, Beijing, Guangdong, Tianjin, Hubei and Chongqing. The power industry is an important participant in these pilots and almost all power companies fall within their scope: more than 150 companies have been involved in transactions through the markets. At present, the *Interim Measures for the Administration of Carbon Discharges Trading* governs the nascent carbon markets, but a nationwide carbon trading scheme was expected to be launched in China in late 2017 [38], though as of going to press no launch date had been confirmed.

According to simulations run by the Institute of Science and Development at the Chinese Academy of Sciences (CASISD), the following observations can be made of China's carbon markets:

- Running costs for new power plants may rise by a small margin. Carbon costs for new-build thermal power plants within the pilot carbon markets could account for up to 25 per cent of their levelised cost of energy (LCOE), assuming that the carbon dioxide price remains below RMB 200/tonne. This would be true for all technologies, including ultra-supercritical and supercritical coal and gas-fired plants.
- Generating companies are being incentivised to actively develop additional renewable energy generation capacity. Nonetheless, the carbon cost caused by remaining thermal generation is still too high to be neglected. Thermal generation needs to be reduced by a further 15–20 per cent to reduce carbon costs to more manageable levels.

- Appropriate governing mechanisms for quota auctions can help reduce emissions effectively and efficiently. The initial allocation of quotas is not expected to affect the carbon markets much when a suitable auction mechanism is put into practice; later trading, however, can be optimised through two-way (or double) auctions. This can prevent a small number of large players from cornering the market and encourage widespread participation and competition.

The carbon trading system is an important part of the government's plans to change the structure of generating costs. Power generation companies will benefit from monitoring their emissions more carefully (as is necessary to ensure they are in compliance with emission allowances) and this will help encourage steps to reduce overall emissions. Measurement is often the first step towards action, and carbon allowances, alongside the carbon markets, will force accurate and timely measurement. Emission forecasts will also begin to feature in long-term industry strategic plans and become a key part of future project investment decisions. As the sophistication of both carbon markets and industry participants improves, further progress can be expected in the continued reduction of carbon emissions.

Energy and Carbon Taxes

Energy taxes are payable by end-users of electricity. Energy taxes have been implemented since the beginning of the twentieth century in many countries in Europe and in the United States, and have helped promote energy efficiency and environmental protection.

In China, there are several types of energy tax:

- value-added tax on energy sales;
- taxes on energy consumption (such as a fuel tax); and
- resource taxes levied on energy project development.

There is, as yet, no unifying single energy tax, though discussions within academic and government circles continue.

In 1992, Denmark became the first country in the world to levy a carbon tax; other countries have since followed suit. In China, the *Environmental Protection Tax Law* was passed in 2016, and it is likely that a carbon tax will be added to the environmental tax system as a supplement to the carbon trading policy after 2020.

With a carbon tax and energy tax levied, it will be possible to stimulate the development and utilisation of new technologies and new forms of renewable energy across the power industry: thermal generation should become more efficient; the fuel mix should again be shifted towards a cleaner and lower carbon mix; and industry, in general, should be incentivised to become cleaner.

Levying both an energy and carbon tax should:

- Reduce the power sector's energy use and carbon emissions, higher tax rates being of greater consequence. While a carbon tax might be expected to deliver a significant shift towards low-carbon fuels in the medium term, it may also be vulnerable to the rebound effect as improved energy efficiency for individual end-users leads to energy used elsewhere, with carbon saved in one area being emitted in another.
- Stimulate the adoption of low-carbon technologies and fuels, making the sector cleaner and more efficient. The different costs of different technologies mean that their adoption is sensitive to the ultimate tax rate.
- Raise costs and reduce employment within the power sector in the short term. In these respects, a carbon tax is expected to have a greater impact than an energy tax. It is estimated by CASISD that by 2020, a carbon tax will reduce employment in the power sector by 9.6 per cent, and generation capacity by 10.6 per cent. If an energy tax is levied instead, the corresponding decreases will be 3.5 per cent and 3.8 per cent, respectively.
- Rising power prices should reduce demand. Along with promoting technological upgrades, reducing demand (demand-side management) is seen as a major pathway to reduced emissions. At low tax rates, demand reduction is probably a bigger factor than technology upgrades, but as taxes rise, technology will play a larger part in reducing emissions.

Delay in introducing these taxes could actually raise overall costs to the industry. Emission reductions will need to be made more deeply and steeply the longer they remain unchecked, raising costs substantially if the same emission targets are to be achieved. It may be better for the Chinese government to implement such taxes as soon as possible, to give enterprises adequate time to adapt and put in place mitigation plans. Carbon and energy taxes can always start low (to gain acceptance) and ratchet up through time to deliver the required environmental results.

UNCERTAINTIES IN THE EFFECTS OF ENVIRONMENTAL REGULATIONS ON THE ELECTRIC POWER SECTOR

At present, the electricity sector is not a true market: power prices in China are not fully liberalised and generation quotas are allocated by the government. With government policy directly affecting the industry in multiple, sometimes conflicting, ways, the overall effect of environmental regulations is not always certain.

Challenges in New Energy Investment

Environmental regulations have accelerated investment in new energy projects by power generating companies and have promoted the optimisation of the power supply fuel mix. But it is worth noting that although the installed capacity of renewable energy has grown remarkably, corresponding levels of actual generation lag behind. In addition to the general nature of intermittency, curtailment of wind and photovoltaic power are serious ongoing problems. Taking GD Power Development Company (GDPD) as an example, in 2015, clean energy (meaning non-fossil energy) accounted for 32.6 per cent of total installed capacity, but curtailment meant that it accounted for only 23.4 per cent of generation.

As stipulated in the *Guidance of the National Energy Administration on the Development of Guiding System on Exploration and Utilisation Targets for Renewable Energy* (2016), by 2020 all generating companies need to have at least 9 per cent of their power generated by non-hydro renewable energy [39]. 55 per cent of China's wind and solar capacity is in the hands of its five largest generating companies; of these five, GDPD has the highest proportion of renewable generation (about 6.2 per cent), while CHC has the lowest (2.5 per cent). In all cases, there is some way to go to reach the 9 per cent target [40].

Wind and solar curtailment is a result of slowing electricity demand growth, excessively rapid development of new generating capacity and insufficient connections through the grid. Increasing R&D investment, particularly into the technical difficulties associated with electricity storage and power transmission, will be critical to long-term reductions in curtailment. In the medium term, a more coordinated approach to the complementary advantages of different fuel types (including coal, LNG, nuclear, renewables and pumped-storage hydro) will help optimise generation. In addition, priority for renewables, green certificates and their trading, proper market reforms and clearer price incentives will all help promote a cleaner power sector. Curtailment is discussed in more detail by Sufang Zhang in Chap. 4.

Challenges for Coal Power Emission Reduction

The two campaigns of 'replacing smaller units with big ones' and 'ultra-low discharge' have helped boost clean coal in China, cutting sulphur dioxide and nitrogen oxides emissions. The campaigns have also had unintended consequences.

Despite China's surplus generating capacity, there continue to be new capacity additions. Of the 15 coal projects cancelled by the National Energy Administration in September 2016, ten were power plants in the 'replacing smaller units with big ones' programme [41]. Explicit constraints need to be imposed on the overall development of coal-power generation; otherwise, the 'replacing smaller units' campaign could have the side effect of actually increasing coal capacity yet further as larger units are constructed. This is resulting in low asset efficiency, poor investment returns and rising emission control costs. In 2015, China increased its generating capacity by 72 GW, of which coal accounted for almost 52 GW: the highest increase since 2009 [2].

Of the 'ultra-low discharge' programme, many of the standards, technical specifications and environmental targets and benefits have yet to be properly worked out. Thus far, the programme has focused on the eastern provinces, but as the upgrade activities head inland to the central and western provinces, new problems could arise. One of these at least lies in enforcement. Upgraded facilities could lead to an increase in energy consumption. Meeting the 2020 deadline for full implementation will be a challenge, requiring both government and business to fully understand the costs and benefits, while continuing to explore more cost-effective ways to meet the environmental regulations.

Challenges for Business Management

Environmental constraints require generating companies to invest in changes to both fuel and equipment, potentially raising costs and reducing profits if prices cannot be raised. Following the actions outlined within the 11th and 12th FYPs, the space for further low-cost energy savings and efficiency gains in coal-fired power plants is largely gone; the next stage is expected to require substantially more upfront investment. At the same time, with the implementation of the current round of electric power market reforms, subsidies will likely be slowly withdrawn, leaving the industry to bear the costs of meeting environmental regulations themselves.

Business management strategies must also consider further issues and risks such as the increased downtime experienced through maintenance schedules for (and unplanned shutdowns of) desulphurisation and denitrification equipment, as well as simply their running costs. Traditional businesses may also find themselves locked out of certain services as priority is given to other groups such as renewables dispatching to the grid or cogeneration plants providing heating. Finding themselves lower down various merit orders, traditional generating companies will face new risk profiles and will need to manage their exposure to this new operating environment.

Clearer planning and investment discipline will also be required in an industry that has to date been able to expand relatively easily. As environmental regulations impose new costs and shift risk profiles, managers will need to become more sophisticated in their planning so as to avoid falling foul of environmental regulation or failing to generate sufficient return on investment decisions. As always, managers will need to balance various costs and alternatives. The value of information—of clear, consistent, timely and transparent data—can, therefore, be expected to rise in tandem with the industry's environmental transformation.

CONCLUSIONS

China's environmental regulations are becoming more and more comprehensive, successfully making the electric power sector cleaner and more efficient.

The country has established a multi-layered policy and legal basis for its long-term ambition to transition to a low-carbon economy built around a low-carbon power sector. FYPs, long-term strategic plans, laws—such as the *Environmental Protection Law of the People's Republic of China* and the *Law of the People's Republic of China on the Prevention and Control of Atmospheric Pollution*—and specific regulations ranging from prioritised dispatch for renewable energy to reduced quotas for coal-fired generators are all contributing to an increasingly effective and stringent framework.

Alongside traditional administrative tools, China has also begun to turn to the market to push its environmental agenda. Taxes, tradable pollution permits, green certificates and carbon markets are all increasing the importance of price signals to both generators and end-users, an important change in a sector that has long been immune to price and with few feedback mechanisms to temper demand or raise efficiency. To a certain extent, then, environmental protection measures can also be placed within the context of the broader market reforms that started in March 2015 (see Chap. 3 for a fuller discussion).

Despite the clear strength of the government's ambition and commitment, China's power sector remains reliant on coal and is a source of much pollution. The growth in renewable capacity has not been matched by the growth in renewable generation, while the whole industry faces economic headwinds given serious overcapacity. Exactly how the government implements and enforces its new environmental protection measures will go a long way to determining their effectiveness, but success will also rely upon the power industry refreshing its business models, properly accounting for environmental effects and engaging proactively with both regulation and consumers.

Beyond the headlines about air quality, there have been headlines about the structural shifts taking place in China's power sector. The environmental regulations outlined in this chapter make it clear that these adjustments are far from finished.

REFERENCES

1. The State Council of PRC (2016) *A Comprehensive Plan for Energy Conservation and Emission Reduction During the 13th Five-Year Plan Period.* http://www.gov.cn/zhengce/content/2017-01/05/content_5156789.htm
2. National Bureau of Statistics of PRC (2017) *National Data.* http://data.stats.gov.cn
3. National Development and Reform Commission (2016) *Plan for Energy Development During the Thirteenth Five-Year Plan Period.* http://www.ndrc.gov.cn/zcfb/zcfbtz/201701/t20170117_835278.html
4. Ministry of Environmental Protection of PRC (2016) *National Environmental Statistics Bulletin.* http://www.zhb.gov.cn/gkml/hbb/qt/201706/t20170605_415442.htm
5. Ministry of Environmental Protection of PRC (2015) *National Environmental Statistics Bulletin.* http://www.zhb.gov.cn/gzfw_13107/hjtj/qghjtjgb/201702/t20170223_397419.shtml

6. BeiJiXingDianLiWang (2017) *Effects of Environmental Regulation on Power Sector.* http://news.bjx.com.cn/html/20131204/477397.shtml
7. Ministry of Environmental Protection of PRC (2014) *Environmental Protection Law of PRC (Amendment).* http://www.zhb.gov.cn/gzfw_13107/zcfg/fl/201605/t20160522_343393.shtml
8. Ministry of Environmental Protection of PRC (2015) *Law of the People's Republic of China on the Prevention and Control of Atmospheric Pollution (Amendment).* http://www.zhb.gov.cn/gzfw_13107/zcfg/fl/201605/t20160522_343394.shtml
9. The National People's Congress of PRC (2016) *Environmental Protection Tax Law.* http://www.npc.gov.cn/npc/xinwen/2016-12/25/content_2004993.htm
10. Xinhuanet (2016) *Outline of the 13th Five-Year Plan for the National Economic and Social Development of the People's Republic of China.* http://news.xinhua-net.com/politics/2016lh/2016-03/17/c_1118366322.htm
11. The State Council of PRC (2017) *Plan for Energy Conservation and Discharge Reduction During the Thirteenth Five-Year Plan Period.* http://www.gov.cn/zhengce/content/2017-01/05/content_5156789.htm
12. The State Council of PRC (2016) *Plan for Environmental Protection During the Thirteenth Five-Year Plan Period.* http://www.gov.cn/zhengce/content/2016-12/05/content_5143290.htm
13. China Power (2016) *Plan for the Development of the Electric Power Sector During the Thirteenth Five-Year Plan Period (2016–2020).* http://www.chinapower.com.cn/focus/20161108/64097.html
14. Ministry of Environmental Protection of PRC (2012) *Emission Standard for Air Pollutants for Thermal Power Plants.* http://kjs.mep.gov.cn/hjbhbz/bzwb/dqhjbh/dqgdwrywrwpfbz/201109/t20110921_217534.htm
15. National Development and Reform Commission (2014) *National Plan in Response to Climatic Changes (2014–2020).* http://www.ndrc.gov.cn/zcfb/zcfbtz/201411/t20141104_642612.html
16. National Energy Administration (2014) *Energy Development Strategy Action Plan (2014–2020).* http://www.nea.gov.cn/2014-12/03/c_133830458.htm
17. Ministry of Environmental Protection of PRC (2003) *Measures for the Administration of the Charging Rates for Pollutant Discharge Fees.* http://www.zhb.gov.cn/gzfw_13107/zcfg/gz/bmgz/gwybmyggz/201605/t20160531_352590.shtml
18. Ministry of Finance of PRC (2015) *Interim Measures for the Administration of Revenue Management for the Sale of Pollution Rights.* http://szs.mof.gov.cn/zhengwuxinxi/zhengcefabu/201507/t20150731_1397067.html
19. Ministry of Environmental Protection of PRC (2016) *Notice on Deploying the Management over the Pollutant Discharge Permit from Elevated Sources in Pilot*

Cities in Beijing, Tianjin and Hebei Province as Well as in the Thermal Power Industry and the Papermaking Industry. http://www.zhb.gov.cn/gkml/hbb/bwj/201701/t20170105_394016.htm

20. Ministry of Environmental Protection of PRC (2016) *The Environmental Impact Assessment Law of the People's Republic of China (2003, Revised in 2016).* http://www.zhb.gov.cn/gzfw_13107/zcfg/fl/201609/t20160927_364752.shtml

21. National Energy Administration (2017) *Notice on the Issue of Renewable Energy Power Green Certificates and Its Voluntary Subscription Trading System.* http://www.nea.gov.cn/2017-02/06/c_136035626.htm

22. The Central People's Government (2014) *Instructional Advice on Implementing Pilot Work on Paid Use and Trading of Emission Permits.* http://www.gov.cn/zhengce/content/2014-08/25/content_9050.htm

23. The Central People's Government (2016) *Implementation Plan for the Permit System for Controlling Pollutants Emission.* http://www.gov.cn/zhengce/content/2016-11/21/content_5135510.htm

24. Ministry of Environmental Protection of PRC (2003) *Notice on Enhancing the Prevention and Control over Sulphur Dioxide Pollutions Caused by Coal-Fired Power Plants (2003).* http://www.zhb.gov.cn/gkml/zj/wj/200910/t20091022_172226.htm

25. National Development and Reform Commission (2013) *Notice of the National Development and Reform Commission on the Additional Standards for Making Adjustments to the Price of Electric Power Generated from Renewable Energies and Proceedings Related to the Prices of Environment-Protecting Electric Power.* http://www.ndrc.gov.cn/zcfb/zcfbtz/201308/t20130830_556008.html

26. National Development and Reform Commission (2015) *Notice on Issues Related to the Implementation of Supporting Policies on the Price of Electric Power Generated by Coal-Fired Power Plants with an Ultra-Low Emission.* http://www.ndrc.gov.cn/zcfb/zcfbtz/201512/t20151209_761936.html

27. The National People's Congress of PRC (2016) *Electricity Law of the People's Republic of China.* http://www.npc.gov.cn/wxzl/gongbao/2015-07/03/content_1942878.htm

28. National Development and Reform Commission (2016) *Measures for the Administration of the Guaranteed Purchase of Electricity Generated by Renewable Energy.* http://www.ndrc.gov.cn/zcfb/zcfbtz/201603/t20160328_796404.html

29. National Development and Reform Commission (2017) *Notice on Releasing Electric Power Generation Plans and Electric Power Use Plans in an Orderly Manner.* http://www.ndrc.gov.cn/gzdt/201704/t20170410_843778.html

30. People's Daily (2017) *China Leads the Development of New Energy Globally.* http://paper.people.com.cn/rmrb/html/2017-01/19/nw.D110000renmrb_20170119_1-24.htm

31. China Electricity Council (2017) *2016–2017 National Power Supply and Demand Analysis and Forecast Report.* http://www.cec.org.cn/yaowen-kuaidi/2017-01-25/164285.html

32. China Electricity Council (2017) *National Electricity Supply and Demand Situation Analysis Forecast Report for the First Half of 2017.* http://www.cec.org.cn/yaowenkuaidi/2017-07-25/171317.html

33. National Development and Reform Commission (2017) *Administrative Measures for Power Demand-Side Management.* http://www.ndrc.gov.cn/fzgggz/jjyx/dzxqcgl/201011/t20101116_381342.html

34. National Development and Reform Commission (2007) *Notice of Several Opinions on Accelerating Close-Down of Small Thermal Power Generating Units.* http://www.ndrc.gov.cn/zcfb/zcfbqt/200701/t20070131_115037.html

35. FengHuangWang (2017) *Shut Down Small Thermal Power and Reduce Carbon Dioxide Emissions.* http://finance.ifeng.com/roll/20101116/2892765.shtml

36. Ministry of Environmental Protection of PRC (2015) *Implementation of Transformation Oriented at Ultra-Low Discharge and Energy Conservation in Coal-Fired Power Plants.* http://www.zhb.gov.cn/gkml/hbb/bwj/201512/t20151215_319170.htm?_sm_au_=iVVR2PCFSksVLj6H

37. China Economic Times (2017) *Control the New Capacity to Promote Coal-Powered Electricity Transformation and Upgrading.* http://www.cet.com.cn/nypd/yw/1953485.shtml

38. National Development and Reform Commission (2014) *Interim Measures for the Administration of Carbon Discharges Trading.* http://qhs.ndrc.gov.cn/zcfg/201412/t20141212_652007.html

39. National Energy Administration (2016) *Guidance of the National Energy Administration on the Development of Guiding System on Exploration and Utilisation Targets for Renewable Energy.* http://zfxxgk.nea.gov.cn/auto87/201603/t20160303_2205.htm

40. BeiJiXingDianLiWang (2016) *Have the Five Major Power Generation Groups Had Enough Renewable Energy Yet?* http://news.bjx.com.cn/html/20160401/721557-2.shtml

41. National Energy Administration (2016) *Notice on Cancelling a Batch of Coal Power Projects that Do Not Have the Approved Construction Conditions.* http://zfxxgk.nea.gov.cn/auto84/201609/t20160923_2300.htm

CHAPTER 6

Financing China's Electricity Sector

Dayong Zhang and Huadong Dai

Abstract China's total investment in the electricity sector during the 12th Five Year Plan (FYP) period reached RMB 3.9 trillion. Equity financing has become increasingly important and other innovative instruments have been developed. Moreover, the booming renewable energy industry has brought new features to the electricity sector. As the 13th FYP further reinforces the importance of renewable energy and the upgrading of China's existing energy system, there will inevitably be further reforms and changes, which will consequentially affect the financing structure of this industry. This chapter begins with a brief introduction to the current state of financing in China's electricity sector, before discussing its future development. The hope is to provide interested readers with a general picture of financing in China's electricity sector.

Keywords Capital structure • Electricity sector • Financial innovation
• Green finance • Market reform

D. Zhang (✉) • H. Dai
Research Institute of Economics and Management, Southwestern University of
Finance and Economics, Chengdu, China

© The Author(s) 2018   133
L. Lester, M. Thomas (eds.), *China's Electricity Sector*,
https://doi.org/10.1007/978-981-10-8192-7_6

Summary

China's total investment in the electricity sector during the 12th Five-Year-Plan (FYP) reached RMB 3.9 trillion. Although debt financing accounts for a quite critical part, especially for major state-owned enterprises, equity has become increasingly important. This growth in equity has resulted from the fact that most firms on the power generation side have become publicly listed. The phenomenal development of China's capital markets, such as its stock, bond and derivative markets, has provided electricity firms with more opportunities to acquire financing resources at lower cost. Financial innovations in the capital markets, such as green bonds, asset-backed securities and carbon financing, have brought new features to the power sector.

Total investment in power generation in China during the 13th FYP is predicted to be RMB 2.84–3.04 trillion; RMB 2.75 trillion is expected to be invested in the grid. As the 13th FYP further reinforces the importance of both renewable energy development and the upgrading of China's existing energy system, there will inevitably be further reforms and changes that will consequently affect the financing structure of the industry. One has to notice that over decades of electricity market reform, the general objective of the Chinese authorities has been to move away from a system of central planning towards more market-driven mechanisms.

China still relies heavily on traditional fossil fuels, especially coal, to generate electricity and the dominance of coal is unlikely to change in the near future. However, renewable electricity will expand and play a more important role as the enormous environmental pressures force the government to shift towards cleaner energy.

Renewable electricity firms are generally smaller, more innovative and have clearly different cost-return structures relative to traditional electricity firms. With the 13th FYP setting out the establishment of a green financing system, renewable energy firms can now benefit from both more favourable government policy and China's quickly expanding financial markets. For example, global green bond issuance had ramped up to around USD 81 billion by 2016, with China alone accounting for 29 per cent of the global total. A nationwide carbon trading system is on its way, which will provide renewable electricity firms with another great opportunity.

Beginning with an introduction to macro-level background information and the current status of financing issues in China's electricity sector, this chapter moves on to discuss the available financial instruments and micro-level company financial structures. Case studies are provided to give the reader some real-life examples of the new financing features in China's electricity sector. In general, this chapter aims to provide information on what has been going on in the electricity sector in China since the 1978 economic reforms. Through a discussion of the status of development at both macro and micro levels, we will look at the challenges and future financing issues faced by the electricity sector in China.

KEY INSIGHTS

- Recent electricity market activities suggest that the involvement of the private sector will continue to be encouraged and market mechanisms continue to be enhanced. The generation sector is relatively more flexible than the power grid, and the renewable power sector will be the most active and dynamic part during this marketisation process. The increasing number of financial instruments available, combined with fast-developing capital markets in China, gives the electricity sector more diversified financing opportunities. Investors are likewise provided with more opportunities to exploit the growth in China's electricity sector.

- The slowing down of China's economic growth has led to lower growth in electricity consumption. The oversupply of electricity, especially of coal-fired generation, can cause significant losses to both the industry and investors. Overinvestment also exists because of agency problems and the market risks in various financial products need to be considered by investors and regulators (especially in the light of the 2008 global financial crisis). While a market-oriented system can better reflect the fair value of electricity and increase efficiency, associated risk exposures will also increase. For example, non-synchronised coal and electricity price movements can bring higher risks to coal-fired power plants.

- China's electricity sector will remain largely under state control and it is unlikely to have significant involvement of international capital. The market-oriented reform and diversification of financing products, however, provide a chance for foreign investors. Given the size of the Chinese electricity market, even a small share can lead to major opportunities. Many new developments and products have international roots, giving international investors a comparative advantage. For example, the carbon trading system largely follows the EU ETS;

there are already decades of experience in third-party electricity trades in the USA, EU and other developed economies. Financial instruments such as asset-backed securities and financial leasing all have a longer history in international markets.

INTRODUCTION

The electricity sector has been crucial to the overall development of China's economy. From earlier stages with power shortages to the most recent oversupply of electricity, China's power system has been shown to be one of the most important factors supporting the country's rapid economic growth [1]. The achievements have been phenomenal, in both the scale and speed of development. According to the World Bank World Development Indicators (WDI), China achieved 100 per cent electrification in 2012; its most comparable close neighbour, India, had only 78.7 per cent of its total population with access to electricity that year.

In July 1955, the Ministry of Power Industry was established as the main regulatory body for China's electricity sector. Alongside China's post-1978 economic reforms, the electricity industry in China has also gone through a series of major changes from a centrally planned system with state monopolies to its recent more market-oriented structure [2]. The Ministry of Power Industry was restructured, replaced and re-established several times before the Ninth People's Congress announced its final disbandment in 1998. The State Power Corporation was established in 1997 with the aim of replacing the Ministry of Power Industry and introducing modern corporate management systems. In 2002, further market reform took place, and the State Power Corporation was split into 11 entities, including two major grid companies: the State Grid Corporation of China and the China Southern Power Grid Company Limited. The other entities comprised five power generation corporations and four other supporting companies. This restructuring, and the establishment of the State Power Regulatory and Management Commission in the following year, marked a new era for China's electricity sector. Similar to the separation of owners and managers in modern corporations, the power industry now had a separation of government authority and ownership from its corporate management.

Although a series of reforms have steadily driven China's power sector towards a more diversified and market-oriented structure that both allows and encourages competition, the state is still the main controlling power.

For example, the National Development and Reform Commission (NDRC) is the main government agency responsible for planning new electricity projects and setting tariffs. Nuclear, natural gas and hydropower assets are mainly state-owned.

As a consequence of these major organisational changes and reforms, a more diversified financing structure in China's electricity sector was introduced. In 1985, the State Council issued the *Provisional Regulation on Encouraging Fund-Raising for Power Construction*, which enabled new classes of investors (such as local governments and even international capital) to enter China's domestic electricity market. This significant change helped push the development of the power sector.

Passed in the Eighth People's Congress in 1995, the *Electricity Law of the People's Republic of China* came into force in April 1996. The first clause of the first chapter states clearly that protecting the lawful rights of electricity investors is one of the most important objectives of the law. The main principle is to provide a legislative guarantee that investors can benefit from their investment in the electricity sector. This legal protection further encouraged domestic and overseas investors to get involved. As a result of these reforms, China's electricity sector expanded rapidly, with an annual growth rate of around 9 per cent during the period from 1985 to 2014. Total electricity production soared from 411 TWh, in 1985, to 5,650 TWh in 2014 [3], by which point China had become the largest electricity producer in the world.

By now, most of the big firms in the power sector have listed on stock markets (as A shares in mainland China or H shares in Hong Kong), which further increases the channels of financing. Deregulation and diversification have brought vitality to the power industry, and investment has grown strongly since 2006. The average annual growth in investment in the electricity sector was over 5 per cent during the 11th and 12th FYP periods (2006–10 and 2011–15). Figure 6.1 plots annual investment and its growth rate.

Table 6.1 provides some detailed investment information in subcategories during this period. A clear slowing down of investment growth is observable after 2010. This slowdown is not entirely surprising: the 2008 global financial crisis caused a global economic recession, from which there has been only a partial recovery in recent years. Given that China used to rely heavily on exports to fuel its rapid economic growth, weaker demand from the global market hurt the 'world's factory' quite

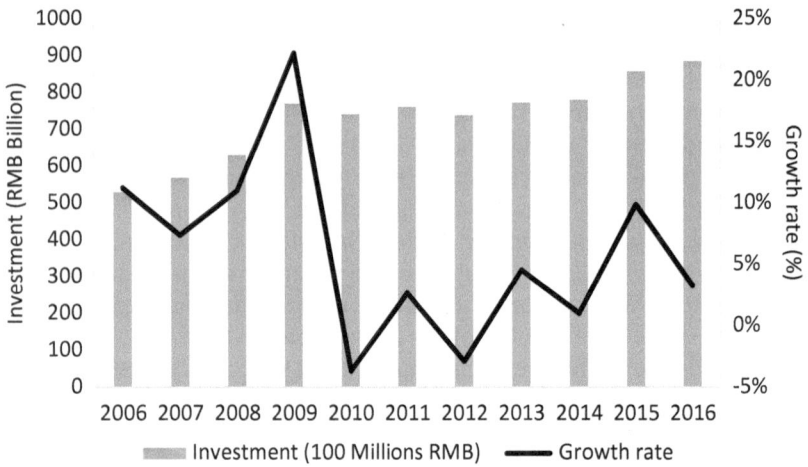

Fig. 6.1 Annual investment in the electricity sector. Bars show annual investment totals on the left-hand axis; the line shows growth in investment flows on the right-hand axis (Source: China Electricity Council [4]; total investment does not include solar or bio-power before 2011)

Table 6.1 Electricity sector investment in the 11th and 12th FYP periods, 2006–15

Year	2006	2007	2008	2009	2010	2011	2012	2013	2014	2015
Generation	317	320	338	378	393	393	373	387	369	394
Hydro	78	86	85	87	82	97	124	122	94	79
Thermal	223	200	168	154	143	113	100	102	115	116
Nuclear	9	16	33	58	65	76	78	66	53	57
Wind	6	17	53	78	104	90	61	65	92	120
Solar	N/A	N/A	N/A	N/A	N/A	16	10	32	15	22
Grid	209	245	290	390	345	369	366	386	412	464
Total	529	568	630	770	742	761	740	773	781	858

Source: China Electricity Council [4]. Investment in solar energy is not included before 2011. Unit: Billion RMB

significantly. The Chinese economy has entered a 'new normal' stage, characterised by a shift from high-speed growth to a medium level of growth. There is also significant industrial overcapacity, which has caused a reduction in the rate of growth of electricity consumption.

Looking at the slowdown in investment at the aggregate level reveals the changing international economic environment and poses questions for the development of China's electricity sector. The sectoral breakdowns, though, can give quite a different story. Over the whole ten-year period, investment in the power grid accounted for 49 per cent of the total, not surprising given China's need to upgrade the whole power system to achieve higher efficiency. The interesting part comes from the power generation side. Although traditional thermal electricity generation still accounts for the highest share, at 20 per cent, it has experienced negative growth for most of the period. In contrast, investment in the renewable sector (excluding hydropower) has grown remarkably. Total cumulative investment in nuclear, wind and solar is equal to 18 per cent; hydropower alone accounts for 13 per cent of total investment. Investment in wind power has grown from RMB 6.3 billion to RMB 1200 billion, an increase of almost 20 times.

This growth in renewable energy is likely to continue. China has committed to peak its carbon emissions by 2030, an announcement reaffirmed in the 2015 Paris UN climate change conference (COP21). This inevitably requires China to adopt more renewable energy in current and future FYPs (i.e. 13th FYP and beyond). This commitment will boost investment in the renewable energy sector significantly.

In general, China will carry on investing heavily in its electricity sector in order to upgrade its power grid and shift more capacity towards renewable energy. As deregulation continues to unfold, financing will become more diversified, especially on the power generation side. The opportunities to investors remain attractive, though, as always, understanding the behaviour of electricity companies is important.

A more market-based system provides higher flexibility for business and investors, but it may also generate higher risks. Firms run the risk of irrational behaviours because of the principal-agent problem. Powerful managers, especially those in major state-owned companies, can have quite different objectives from shareholders and owners. The investment decisions they make can hurt the interests of investors. For example, Zhang et al. [5] studied this problem empirically based on the listed energy-related companies in China. They found that managers tended to invest in low-return investment projects that brought themselves benefits, instead of distributing profit to shareholders. This is also relevant for regulators: understanding the status of, and problems with, financing in the electricity sector can assist them in designing policies that will support the sustainable development of the sector.

The remaining parts of this chapter will go over the available financing options of electricity companies, the status of the electricity sector's capital structure and whether capital structure matters based on standard corporate finance theory and some recent studies. Two case studies will then be used to show the opportunities and challenges for investors in the electricity sector when financial innovation takes place. Finally, a forward-looking perspective is taken to look at electricity sector financing in the light of the 13th FYP and possible future developments and challenges. The last section concludes briefly.

FINANCING IN CHINA'S ELECTRICITY SECTOR

While the last section provided a general macro-level overview of the development of China's electricity sector and its associated investment trends, this part will look at financing issues in the electricity sector at the firm level. Using listed firm data, we gather information on the capital structure of the electricity sector, introduce the financial instruments available and discuss the question of whether capital structure matters to electricity companies.

How to finance an investment project is the first and the most important question in corporate financial management. Mainstream modern corporate finance literature, such as the Modigliani and Miller Theorem [6], starts with the idea that capital structure is not relevant. It states that a firm's performance is essentially related to the value of its underlying investment projects, not with how these projects are financed. The theory is based on a series of strong assumptions that are not met in reality: corporate tax shields, the cost of financial distress and imperfect capital markets can all cause significant distortions invalidating the theorem. Empirically, international evidence on capital financing decisions leads to the pecking order theory of corporate investment, which suggests that firms prefer internal funds over debt financing and debt financing over equity financing.

Capital Structure of Electricity Firms

To understand the significance of capital structure in China's electricity sector, we start with a general description of the current status using a set of listed firms. The sector's innate monopolistic nature, and history of central planning, has meant that the major corporations are all state-owned.

Investment by the state, and state-owned financial intermediaries, plays a very important role and results in a higher debt to asset ratio than the average for companies listing A shares on Chinese stock markets. However, compared with developed economies, such as the USA, the level of leverage in the Chinese electricity sector (in terms of listed firms in the A share market) is still relatively low (see Table 6.2 for debt to asset ratios between 2006 and 2015).

In recent years, deeper market system reforms and the faster development of capital markets have brought significant changes to the general picture. It is noticeable that the debt to asset ratio differs significantly across sub-sectors. The power grid used to be the one with the highest leverage but since 2009 to 2011 it has been overtaken by the renewable sector and by thermal energy firms. All electricity firms showed a significant increase in debt to asset ratios during the 2008 global financial crisis period. The Shanghai Composite Index (for A shares) fell from 6214 (16/10/2007) to 1660 (19/09/2008); while the value of debt was not affected, the fall in the stock market reduced the value of company assets, raising the debt to asset ratio.

Table 6.2 Debt to asset ratios (shown as percentages) for listed electricity firms between 2006 and 2015

Divisions	2006	2007	2008	2009	2010	2011	2012	2013	2014	2015
Hydro	56.9	60.09	59.94	61.87	50.77	51.65	51.79	52.46	49.67	44.67
Thermal	52.79	53.77	63.74	66.79	67.1	69.95	69.43	63.42	61.48	60.17
Renewables	58.32	59.49	57.97	62.97	61.19	63.15	62.07	63.86	71.94	64.19
Power Grid	61.82	61.52	63.95	64.62	61.56	58.84	58.32	57.92	54.2	53.66
Total	57.46	58.72	61.4	64.06	60.16	60.9	60.4	59.42	59.32	55.67
US electricity firms (average)	62.46	62.11	66.72	63.58	65.7	66.85	69.72	65.87	68.44	67.92
Chinese market average (A shares)	51.76	50.41	54.14	44.1	56.44	42.75	42.08	43.2	40.63	39.17

Source: Wind Finance Database [7] and the authors' calculation. The Chinese market average numbers exclude financial companies

Financial Instruments

The pecking order theory of corporate finance separates financing sources into internal financing and external financing. Internal financing is mainly through retained earnings, whereas external financing includes debt financing and equity financing. Traditionally, firms can borrow from commercial banks or policy banks (such as the China Development Bank, one of the main credit suppliers to major infrastructure construction projects in China). They can also list through an initial public offering (IPO) on the stock exchanges or acquire additional capital through seasoned offerings.

With the development of capital markets and financial innovation, firms can now choose from a wider range of financial instruments, though they can still generally be classified into the two main categories:

- Internal financing
  - Internal capital (i.e. retained earnings)
- External financing
  - Equity financing (IPO, seasoned offering)
  - Bank credit (loans from commercial /policy banks)
  - Bonds (general bonds, convertible bonds, green bonds)
  - Financial leases
  - Trusts
  - Asset-Backed Securitisation (ABS)

Despite this variety, China's electricity sector still mainly relies on internal financing, equity financing and bank loans. The development of the stock market in China in recent years has brought electricity companies new opportunities. Not only can big state-owned enterprises get listed on the stock exchanges, small- and medium-sized private firms can also get financing through stock markets. For example, during 2013–16, the 69 publicly listed Chinese electricity companies raised RMB 154 billion from the equity markets, an average of RMB 39 billion each year.

While firms can enjoy the benefits of an expanding stock market and a growing commercial banking industry, they are now also able to choose from a wider range of alternative financing instruments.

Bond Financing

Among all other instruments, bond financing has become an important supplement to bank financing. It is still in an early stage relative to developed economies: domestic bond financing in China for the period 2000–13 accounted for only a small fraction of total debt financing [8]. The ratio of bond financing over bank financing in China is 19.2; in Brazil, it is 46.67 and in South Africa, 24.07.

The corporate bond market is also relatively underdeveloped [9]. The total value of outstanding corporate bonds in China is equal to 14 per cent of GDP; it is 31 per cent for Singapore. Despite its underlying problems, the corporate bond market has been growing. According to S&P Dow Jones Indices, corporate bonds have increased from 10 per cent of the total value of outstanding bonds in 2007 to 33 per cent in 2014 (the rest being government bonds).

Compared to the general corporate bond market in China, the electricity sector tends to use relatively more bond financing. As data from Wind Finance show, in December 2016, the total value of outstanding bonds for China's electricity firms amounted to RMB 1.6 trillion, of which RMB 1.05 trillion was issued by the power generation companies. The offshore bond market in the electricity sector has also grown strongly: by 2016 the total value of outstanding offshore bonds had reached RMB 17.3 billion; there were an additional USD 9.4 billion in US dollar-denominated offshore bonds [7].

The development of the bond market has clear links with the rise of green bonds. Similar to regular bonds, green bonds are essentially a fixed income security, but they differ from others through their commitment to exclusively financing or re-financing green projects [10]. The green bond market emerged in 2007 when the European Investment Bank (EIB) issued Euro 600 million in green bonds. This was followed by the issuance of a corporate green bond in 2013, which triggered further active participation by the private sector. According to the China Green Bond Market 2016 report [11], global green bond issuance had ramped up to almost USD 81 billion by 2016, with almost half of these as corporate bonds. China alone accounts for 29 per cent of global green bond issuance and is the largest issuer of green bonds in the world. This flourishing of the green bond market is consistent with the strategic reform of the electricity sector in China. Following the Paris Agreement, China urgently needs to invest in the renewable sector; the rapid expansion of the green bonds market can help in this regard.

Internationally, the Green Bond Principles [10] set out voluntary guidelines to promote the integrity of the green label. A set of standards for climate bonds, external reviews and green ratings (e.g. Moody's Green Bonds Assessment) have also emerged in the green bond market. The green bond market in China, however, is regulated by the government. Before a green bond can be issued, it must go through inspection by a series of regulatory authorities such as the People's Bank of China, the NDRC and the China Securities Regulatory Commission (CSRC). The China Green Bond Market report [11] summarises these regulatory guidelines into five key points covering eligibility, management of proceeds, information disclosure, verification and policy incentives.

Financial Leases

Financial leases (also known as capital leases) are 'a contract entitling a renter to a temporary use of an asset' [12]. A finance company is typically the legal owner of the asset for the duration of the lease, while the renter (or lessee) has the operating rights over the asset. The lessee needs to pay a series of rental instalments for the operation of the asset and has the option to acquire ownership of the asset.

In order to facilitate international economic collaboration, and the better use of international capital, technology and equipment after the opening-up policy and economic reform in China, China International Trust & Investment Corporation (or CITIC group) was established in 1979, and leasing contracts were first introduced to China. However, the domestic leasing industry only started to expand after 2007. The leasing business has since experienced explosive growth and increased from RMB 8 billion in 2006 to RMB 930 billion in 2011, of which RMB 390 billion was in the form of financial leases. The total outstanding value of lease contracts in 2015 was RMB 3.98 trillion. In September 2015, the State Council announced guidance on speeding up the development of the leasing industry. It aims to establish the legal environment and regulatory support needed by the leasing industry, and to encourage local government to provide financing support to leasing companies. We could expect to see the leasing industry maintaining a high rate of growth over the entire 13th FYP period.

Asset-Backed Securitisation (ABS)

An ABS exchanges the future cash flow of a pooled asset, such as the income from selling the electricity of a wind farm, for an upfront cash payment. An ABS is essentially the same as a Mortgage Backed Security (MBS), which became well-known to people because of the 2008 US subprime crisis. The underlying assets of the securitisation process are normally illiquid and cannot be sold on their own. Pooling them together into a 'financial instrument' enables the owners to sell them to general investors.

The first electricity securitisation in China was initiated in 2004 and approved by the CSRC in April 2006; it raised total capital of RMB 2 billion for the Huaneng Lancangjiang Hydropower Return Special-Purpose Asset Management Scheme. This successful start [13] has set a good example for other electricity companies to follow to access cheap finance and improve their financial structure. The development of ABS within China, however, raises clear issues surrounding regulatory uncertainty, legal restrictions and risk (such as strict and incomplete regulation, difficulties around taking assets off balance sheets and low liquidity). Its development was also affected significantly by the 2008 global financial crisis.

Nevertheless, securitisation remains one of the main directions of electricity sector reform. Since 2015, China's central government has issued a series of policy documents to support further reform of state-owned electricity enterprises. Sina Finance [14] reports that the asset securitisation rates of China's top five power generation groups have reached 42 per cent (Huaneng), 56 per cent (Datang), 48 per cent (Huadian), 57 per cent (Guodian) and 28 per cent (State Power Investment Corporation). In contrast, grid companies such as the State Grid Corporation of China and China South Power Grid have only very small shares. With further policy support, such as further marketisation of these state-owned grid companies, a higher level of securitisation across the entire electricity sector can be expected.

Does Capital Structure Matter? A Discussion Based on China's Renewable Energy Firms

The numbers reported in Table 6.2 show that China's renewable energy sector has had a very dynamic capital structure. Its debt to asset ratio was as low as 58.32 per cent in 2006 and as high as 71.94 per cent in 2014. A question that may naturally arise is whether capital structure actually matters. In other words, to what extent does the choice of financing affect the performance of these companies?

According to traditional corporate finance theories, such as Modigliani and Miller's [6] capital structure irrelevance theorem, a firm's performance is essentially related to the value of its underlying investment projects, not with how these projects are financed. This 'pie' theory simply states that the size of a pie is not affected by how you cut it. The validity of this theory is based on a series of strong assumptions that are not often met in reality. For example, interest payments are exempt from taxation, which generates tax benefits for companies that borrow. This has led to the concept of a corporate tax shield. In other words, a company is better off with higher leverage as it brings greater tax benefits. Higher levels of leverage, however, can increase risks for a company and can cause financial distress, which rules out the possibility of having 100 per cent debt financing. Capital markets are also not perfect, which may bring more complicated distortions. In general, a firm can look for an optimal capital structure that maximises its market value.

Empirical studies that seek to answer the above question are well-established, but the results are mixed and vary by country studied, samples chosen and methods used. Past studies have normally been based on a comprehensive sample of listed companies; few studies have studied just the electricity sector or just the renewable energy sector. The discussion in this part is mainly based on a recent study by Zhang et al. (ZCZ for short) [15] and, to our knowledge, it is the only relevant empirical study that allows us to address the question of whether capital structure matters in China's electricity sector and in its renewable sector in particular.

Using return on assets (ROA) as the dependent variable and various capital structure measures as explanatory variables, this study finds that capital structure can have a statistically significant impact on firms' profitability. The sample used in ZCZ's empirical study is an unbalanced panel data from 2001 to 2013; included in their regression is not only the total debt to asset ratio but also more detailed sub-categories such as corporate bonds, commercial credit and loans with different maturities (short term vs. long term).

The empirical results are quite comprehensive: their samples include upstream equipment producers and downstream power generation firms. For the purposes of this discussion, it is the results for the latter sample that matter more. In summary, the empirical results from ZCZ show that higher debt to asset ratios can improve the performance of renewable electricity firms. Further, firms that have relatively more long-term debt and that use more bond financing and commercial credit are generally better performing [15]. Here we are not trying to overemphasise the results reported in ZCZ, but to suggest for interested readers/investors that capital structure (in other words, how firms finance their investment) matters in terms of company performance. It is important to look into the details of a firm's balance sheet.

CASE STUDIES

In light of the previous discussion on the development of new financial instruments, this section goes one step further to look at the firm level and uses two case studies to illustrate new features of electricity firm financing in China. Both firms are in the renewable energy sector. Longyuan Electric Power Group is the largest wind power supplier in China and an example of a successful player in the green bond market (especially using offshore green corporate bonds). Shenzhen Energy successfully issued China's first solar power ABS project in 2016.

Longyuan Electric Power Group

Longyuan Electric Power Group was established in 1993 and was originally affiliated with China's National Department of Energy. After a series of consolidations and reforms, the company officially became the China Longyuan Power Group Corporation Limited on 9 July 2009. In December 2009, the company listed on the Hong Kong Stock Exchange. The IPO raised HKD 20 billion for the company. The company carries out the design, development, construction, management and operation of wind farms. By the end of 2016, it had a total installed capacity of 19.5 GW, with a total wind power capacity of 17.4 GW, 1.87 GW of thermal power and 0.23 GW of other renewable energy, making it the world's biggest wind power supplier. It also offers consulting, repair, maintenance, training and other professional services for wind farms.

Over the period 2006–15, the company experienced rapid growth: its total assets were worth RMB 14.55 billion in 2006 but RMB 133.47 billion ten years later in 2016. Its debt to asset ratio was generally higher than the industry average (relative to other renewable electricity firms, Table 6.2) before its IPO in 2009, peaking at 80.38 per cent in 2008 (partially due to the 2008 global financial crisis). The ratio fell and levelled off towards the renewable industry average afterwards. In 2016, Moody's credit rating of Longyuan was A3.

The really exciting feature of this company is the way it has obtained financing through corporate bonds. After its IPO, Longyuan issued RMB 7.2 billion in long-term bonds, at an interest rate of 4.8 per cent. In recent years, large-scale issuance of corporate bonds has allowed it to engage in further business expansion at relatively low cost. In 2016, for example, it issued RMB 52.5 billion in short-term bonds, at an interest rate of 2.69 per cent. The total capital raised since listing has exceeded RMB 100 billion.

Moreover, Longyuan successfully raised over USD 1 billion using foreign currency denominated bonds. It is not entirely surprising that the global green bond market has been expanding in recent years [16], the majority of green bonds stem from renewable sector development in China. Although green bonds constitute just a small part of this firm's capital, the development of the international green bonds market introduces a new and reasonably cheap financing channel for renewable energy firms in China.

The renewable energy sector is the key to China's future development and its financial structure is quite dynamic for the power sector. It is not necessary that all renewable energy firms follow Longyuan's example, but this case study demonstrates, at least partially, that the ability to exploit financial instruments and make use of vibrant capital markets is important to the development of electricity firms in China.

Shenzhen Energy ABS Project

The Shenzhen Energy Group Company Ltd (SEC) was established in June 1991 and was listed on the Shenzhen Stock Exchange (SZSE) in September 1993. It was the first electricity firm listed on the SZSE. In 2016, SEC's total assets were worth RMB 60.8 billion. It entered the renewable energy sector in 2007; by the end of 2016, the percentage of its revenue coming from clean energy had risen to 58 per cent. This case study is included not because of SEC's successful transition from traditional energy to renewable energy, but because of its adoption of asset-backed securities as a source of financing.

On 21 January 2016, the Shenzhen Energy-Nanjing Solar PV ABS Project (SNSP) was approved by the authorities and became China's very first solar PV ABS project. The SNSP project was an RMB 1 billion (about 55 per cent of total investment) asset-backed security with ten prime tranches and one subprime tranche of RMB 20 million.

The underlying assets of the ABS were the solar power plants controlled by Shenzhen Energy Nanjing Co. Ltd., a wholly owned subsidiary of SEC; the ABS was guaranteed by SEC. Shenzhen Energy Nanjing is also the holder of the subprime class securities. Bank of China (BoC), Shenzhen Branch, and BoC International Limited are the trust bank and project manager. CCXI, a member of Moody's, provides credit rating services. The ABS is traded on the Shenzhen Stock Exchange.

Solar power plants are eligible for government subsidies, access to the power grid and have stable cash flows. Through ABS, it is feasible to utilise future cash flows to finance new investment projects now and this flexibility can significantly enhance the development of the sector. The successful launching of the SNSP project is an important indicator of future development. In this case, with the ABS product's reasonably short maturity (six months to five years) and the low debt ratio of the issuing company (55 per cent), the risks for investors are generally low.

The SEC ABS project sets up a good example for other renewable power firms, though there are still a couple of issues. First, investors and regulators need to be aware of the existence of fundamental risk: subsidies from government and access to grid are not necessarily guaranteed long term. An ABS with longer maturities may have problems. Second, for SEC and other firms, the cost of capital is still quite high: the expected return on the six-month top prime class is 3.8 per cent.

Although nine years have passed since the 2008 global financial crisis, the damage ABS and other financial innovations in the derivatives market brought to both the financial industry and the global economy still have significant influence. Both regulators and investors seem to be quite conservative in their attitude to the development of the ABS market. Nonetheless, it is a developing instrument, perhaps especially useful for renewable energy firms. In 2016, Golden Wind Tech., another renewable energy company in China, issued an RMB 1.275 billion wind farm ABS. The future of this financing channel looks optimistic.

FUTURE DEVELOPMENTS AND CHALLENGES IN ELECTRICITY SECTOR FINANCING

Slowing Down of the Chinese Economy and Oversupply of Electricity

In recent years, economic growth in China has slowed down from the double-digit numbers seen in previous decades. A 'new normal' state of economic development has been officially acknowledged. As a consequence of this slowing down, there is an oversupply of electricity [17], especially in the thermal power generation sector. Meanwhile, China has made a firm commitment to green development, requiring over 20 per cent of total energy consumed to come from renewable energy sources by 2030, which will cause a further redundancy of thermal power.

According to the China Electricity Council [4], the growth rate of total electricity consumption reached a historic low (since 1998) of 3.8 per cent in 2014; a further slowdown for 2015 is expected from their forecast. The annual utilisation hours for thermal power plants have already fallen to 4706 (out of a maximum of 8760) in 2014: another historic low. Yuan et al. [17] estimated that the rational capacity of coal power is 960 GW given an average of 4.2 per cent annual electricity demand growth during the 13th FYP period. However, Wan and Wu [18] estimate that coal power capacity in China will reach over 1100 GW by 2020, and 1350 GW by 2030, under a business-as-usual (BAU) projection. Yuan et al. [17] conclude, based on the above information, that overcapacity in the coal power sector is severe and that there is a clear investment bubble (overinvestment). Similar estimates suggest that planners need to re-evaluate the need for new generation capacity to avoid further overinvestment [19].

According to the 13th FYP, the planned total investment in power generation sector is around RMB 3 trillion, of which 19 per cent will go to the coal power sector. Yuan et al. [17] estimate that around RMB 700 billion of investment would not be recoverable and that the utilisation hours for an average coal power plant could drop to just 3800.

The problem of overinvestment may also come from irrational behaviour by individual companies, which can also be applied to the renewable energy sector [15]. Managers may invest in unprofitable projects to expand their power but to the detriment of investors.

Further Market-Driven Reforms

China's National Development and Reform Commission (NDRC) issued *Document Number 9* on electricity reform in 2015. It marked a fundamental step forward and is an important breakthrough in the marketisation of the electricity industry. One major change will be to allow third-party electricity dealers (between generators and end users) to participate in electricity trading. In Guangdong province, where these reforms are most advanced, third-party dealers can bid in auctions. During the first round of reform (in 2016), 13 dealerships were granted the right to trade electricity in the market; by the end of June 2017, 322 dealers had entered the Guangdong electricity market, the largest number of any province in China.

The objective of this reform is to allow demand and supply to affect the electricity price and improve efficiency in the electricity market. Market analysts generally suggest that the overall third-party dealer market is expected to grow to RMB 100–200 billion. In addition, while the grid remains under strict government supervision, power generation will be opened up, and is likely to embrace new systems of risk management and investment/financing decisions.

Large-Scale Investment Demand

In November 2016, China's National Energy Administration (NEA) issued the 13th FYP for the development of the electricity sector. As policy guidance, it aims to increase total installed capacity to 2000 GW, with an annual growth rate of 5.5 per cent. Total electricity consumption will rise to between 6800 and 7200 TWh (an annual increase of between 3.6 and 4.8 per cent). The focus is twofold: first an expansion of renewable energy in generation, and second an enhancement of the power grids. Total investment during the 13th FYP period is predicted to be RMB 2.84–3.04 trillion in generation and RMB 2.75 trillion in transmission and distribution.

In order to raise the required capital at a low cost, the power sector will need a more diversified financing structure and a more innovative market-oriented system. Of course, bank loans will still play a major role, but other forms of financing instruments will gain in scale and importance, and bring opportunities for both power companies and investors.

Establishing a Green Financing System

In 2015, the People's Bank of China [20] estimated that at least RMB 2 trillion would be invested in the green sector over the next five years, of which 85 per cent would come from non-government sources. The 13th FYP proposed establishing a green financing system. China is already the world's largest green bond issuer, but this is still insufficient to support its ambitious plan to peak carbon emissions by 2030 and increase the renewable energy share to 20 per cent. Further policies supporting green financing are expected to be seen in the near future.

To cope with environmental pressures and to reduce carbon emissions, China followed the example of the EU Emissions Trading System (ETS) and set up seven pilot emission trading programmes in 2013. The NDRC plans to launch a nationwide carbon trading market, though an official start date has yet to be confirmed. There are currently more than 2000 firms involved in the pilot programmes, with a total trading volume of 160 million tons. In May 2017, the total trading volume had reached RMB 3.7 billion.

The development of carbon trading in China is of great relevance and importance to the power generation sector. It will also likely drive the development of carbon finance. As allowances and green certificates can be traded through the ETS market, it provides a chance for participating firms to exploit new potential financing options. For example, firms with advanced technologies may have lower levels of emission and can therefore sell their excess allowances to acquire additional financing resources. Of course, the current China ETS is still at its R&D stage [21]: overallocation of allowances, low liquidity and trading, enforcement and compliance remain some (though not all) of the challenges policymakers need to consider as the programme develops and expands.

CONCLUSION

The old image of a wholly state-owned, state-run and state-financed power sector in China is no longer as accurate as it once may have been. Through successive reforms, the power sector is changing. While still largely state-owned and largely closed to international finance, the private sector, private investors and private capital are all becoming more important.

Following the listing of many of China's companies, and the evolution of more active and sophisticated capital markets, innovative forms of capital have begun to emerge in the electricity sector, especially in renewable energy. Asset-backed securities, green bonds, financial leases and corporate bonds have all become more common over the last five to ten years.

The influx of private money has also changed the forces acting on the industry. There is now greater pressure facing companies and their managers to improve financial returns. Investors are looking for a return on investment, something that China's chronic overcapacity has depressed.

This chapter has provided a general review of the financing status of China's electricity sector and discussed recent developments. Based on a brief summary of the historical background, we have described a continuing trend towards greater market-oriented policies, affecting both the power sector directly and its financing opportunities. The financing structure of the electricity sector has become more diversified and innovative, bringing new opportunities to investors both domestic and international.

REFERENCES

1. Yuan J, Zhao C, Yu S & Hu Z (2007) 'Electricity Consumption and Economic Growth in China: Cointegration and Co-feature Analysis', *Energy Economics,* 29, 1179–1191.
2. Ngan H W (2010) 'Electricity Regulation and Electricity Market Reforms in China', *Energy Policy,* 38, 2142–2148.
3. CEInet Statistics Database. http://www.cei.gov.cn/
4. CEC (2015) *The Current Status and Prospect of China's Power Industry* (Beijing: China Electricity Council).
5. Zhang D, Cao H, Dickinson D G & Kutan A M (2016) 'Free Cash Flows and Overinvestment: Further Evidence from Chinese Energy Firms', *Energy Economics,* 58, 116–124.
6. Modigliani F & Miller M (1958) 'The Cost of Capital, Corporation Finance and the Theory of Investment', *American Economic Review,* 48, 261–297.
7. Wind Finance Database. http://www.wind.com.cn/
8. Goodell J W & Goyal A (2017) 'What Determines Debt Structure in Emerging Markets: Transaction Costs or Public Monitoring?' *International Review of Financial Analysis* (in press).
9. Hsu S (2014) 'China's Corporate Bond Market: Dying Young?' *The Diplomat.* http://thediplomat.com/2014/07/chinas-corporate-bond-market-dying-young/
10. ICMA (2015) 'Green Bond Principles: Voluntary Process Guidelines for Issuing Green Bond.' March.
11. China Green Bonds Market (2016) https://www.climatebonds.net/files/reports/sotm-2016-a4-en.pdf
12. Investopedia. http://www.investopedia.com/terms/c/capitallease.asp
13. Liu W, Wang J H, Xie J & Song C (2007) 'Electricity Securitization in China', *Energy,* 32, 1886–1895.
14. Sina Finance (2016) *Speeding Up Asset Securitization for Five Power Generation Groups.* http://finance.sina.com.cn/roll/2016-12-07/doc-ifxyicnf1781399.shtml (in Chinese).
15. Zhang D, Cao H & Zou P (2016) 'Exuberance in China's Renewable Energy Investment: Rationality, Capital Structure and Implications with Firm Level Evidence', *Energy Policy,* 95, 468–478
16. Ng T H & Tao J Y (2016) 'Bond Financing for Renewable Energy in Asia', *Energy Policy,* 95, 509–517.
17. Yuan J, Li P, Wang Y, Liu Q, Shen X, Zhang K & Dong L (2016) 'Coal Power Overcapacity and Investment Bubble in China During 2015–2020', *Energy Policy,* 97, 136–144.
18. Wan J & Wu Y (2015) *The Investment Outlook in Electric Power Sector During the 13th FYP Period.* http://www.in-en.com/article/html/energy-2232448.shtml

19. Lin J, He G & Yuan A (2016) 'Economic Rebalancing and Electricity Demand in China', *The Electricity Journal,* 29, 48–54.
20. PBC (2015) 'Establishing China's Green Financial System.' *People's Bank of China Working Paper* No. 2015/7. (in Chinese).
21. Swartz J (2016) 'China's National Emissions Trading System: Implications for Carbon Markets and Trade.' *ICTSD Global Platform on Climate Change, Trade and Sustainable Energy; Climate Change Architecture Series*; Issue Paper No. 6; International Centre for Trade and Sustainable Development, Geneva, Switzerland, www.ictsd.org

Part III

CHAPTER 7

The Last Word: Chinese Exceptionalism

Mike Thomas

Abstract China's challenges are daunting because of their scale and complexity but they are not necessarily unique in their causes or fundamental nature. Nor are the possible solutions open to China unique: the types of regulatory frameworks and market-based mechanisms are limited and have been tried elsewhere. Approvals and cost recovery can be linked to demonstrable need; price-setting and dispatch mechanisms can be made more transparent. As long as investment risks in China remain opaque, private sector involvement is likely to be limited. Even without private sector participation, overall power sector and macroeconomic efficiency will benefit from reform, especially now that excess capacity is unlikely to be mopped up through rapid demand growth. Other countries provide examples of challenges and solutions; exactly what China will do remains unclear.

Keywords Market reform • Regulatory compromise • Resource imbalance • Investor risk • Energy transition

M. Thomas (✉)
The Lantau Group, Kwai Fung, Hong Kong

China has the world's largest population, the largest car market and the largest (and fastest) high-speed rail network. It has even, by some reports, the most billionaires in the world. It's big and, in terms of both its economic and geopolitical clout, getting bigger. Combined with the length of its history and the speed of its recent economic growth, an image of emerging Chinese exceptionalism appears. China is a civilisation-state with its own economic playbook.

Size can change the number and intensity of problems, but there are still only so many basic types and often even fewer ways to solve them. There may be too much of something because a process or incentive fails to adjust, or a limiting directive is not given or received. There may be the wrong type of something because of a failure to differentiate. There may be a shortage of something because of insufficient incentives, a process that is too slow, or some bottleneck or constraint that has been ignored or underestimated. Economic and political systems differ in the ways they respond to this common suite of problems through price signals, approval processes, regulatory standards and command and control techniques.

China's electricity system has overcome an extraordinary mix of challenges in keeping up with, leading, and supporting the country's economic development, but the process has been anything but orderly or smooth.

- Development has required massive investment, yet China neither anticipated the recent slowdown in electricity demand growth nor has been able to respond effectively to it. The resulting excess capacity is enough to power half of Europe. Excess capacity is hardly new, but the extent of potential stranded costs and the implications for China's economy of mishandling any energy transition (by either raising prices too quickly or withdrawing implicit banking subsidies to the sector too quickly) are of a different scale.
- Once synonymous with air pollution, China is increasingly looked to as a leader in renewable energy and one of the most important participants in global climate accords. But China starts with a power system that generates over 70 per cent of its electricity from coal. Most mature economies face the upcoming energy transition from a position of having a mix of ageing infrastructure and new opportunities, meaning that it is possible to retire older capacity while bringing in new technologies. In contrast, China's power system has largely been built in the last 20 years, with most of that in the last decade.

No country in the world, not even China, is in the fiscal position of being able to afford to shut down and replace *brand new* capacity on the scale required for a rapid low-carbon transition.

In China as elsewhere, the problem of imbalance is a problem of related rates: some things are happening too fast while some things are not happening fast enough. Excess capacity, environmental challenges, resource and locational imbalances, transmission bottlenecks, unsustainable cross-subsidies; they have all been seen before elsewhere. When viewed this way, Chinese exceptionalism is more about the scale of challenges than about the nature of their solutions.

As one peers behind the veil, China's power sector reveals as many similarities with the rest of the world as it does differences. The United States built infrastructure to keep up with electricity demand growth of 7 per cent or more per annum for decades. Singapore suffers from overcapacity and its generating companies are stuck in drawn-out financial pain. America and Germany have both experienced curtailment, even if reports suggest at a lesser rate than in China. London was famous for its coal-fuelled 'pea-souper' smog before the UK Clean Air Act of 1956; smog was legendary in Los Angeles in the 1960s and 1970s, around the same time that the Cuyahoga River in Ohio caught fire because of pollution; Delhi today has air pollution that may be even worse than Beijing's. Even if the route to the problem is unique to each country, and even if the exact nature of the problem displays national differences, the overall challenge is the same. How do countries, economies and societies transition to a low-pollution, low-carbon power sector while balancing the interests of legacy investors, potential new investors, consumers (including those facing energy poverty), policy makers, and taxpayers?

What options are open to China? The most obvious is to change the regulatory process for project approvals and operationalisation. Instead of simply banning (or greenlighting) investment projects according to ministry politics, forecasts and whims, project cost recovery (and by extension overall economics) could be tied to approvals linked to clear, objective and transparent needs. If investors understood that approvals and cost recovery depended upon demonstrable need it would help send the right signals, encouraging investment in times of rapid growth and choking it back when there was little need for expansion. Such approaches would also keep power with the regulators and planners, but would necessitate a different set of attitudes to what has prevailed before. The models for this approach have been in place for a hundred years, used across a wide variety of political and economic systems and, when implemented consistently, have supported economic growth of the highest order. China need only look within itself: Hong Kong's regulatory approach provides an example. By no means are such approaches without flaws—often obvious ones—but they are more inherently self-correcting than what has gone before in China. The biggest challenge in China compared to these other markets is that China's political jurisdictions do not make very good candidates for planning or organising electricity systems. Resources and the loads that must be served are seldom in the same political jurisdiction. Far more so than any other place we know of, China cannot balance demand and supply efficiently as a series of provincial 'islands' with only weak interconnection. Interconnection, multi-regional planning and operational optimisation are key to achieving the best outcomes over time—and these are the challenges that are most difficult to sort out using local or even regional political tools. Even 'simple' regulatory models face difficulties in that one must first agree where the locus of regulatory approval authority should reside.

It is often overlooked that China already operates the largest electricity 'market' in the world in the sense that investors take quite fundamental supply- and demand-related risks when making investments, none of which is protected by long-term regulatory compacts or long-term power purchase agreements or protected by any level of financial certainty. In that respect, China's power sector investment environment looks much more like a competitive 'merchant' market than does Vietnam's or Indonesia's, where investments depend on securing long-term power purchase agreements that are able to compensate investors no matter what level of dispatch is required: a contractual instrument that has not been available in China. Poor transparency and a lack of available contractual protections undermine private sector interest, and therefore participation, in China's power system. Instead, government-linked generating companies have been the major investors, albeit without regard to changing market conditions. Ironically, China has combined a market full of investment risks with stakeholders who are well-placed to ignore them completely. A different balance is needed. Getting behaviours and decision-making better attuned to prevailing risks is key to the success of any regulatory or market-based reforms. If China wants private sector investors to share risk with state-backed companies, then a great deal has to change, beginning with price-setting and dispatch transparency.

Accommodating private sector attitudes to risk is optional, but preventing the power sector from eating up resources that China could more effectively use to accomplish other objectives may not be. When the system was growing so fast that it did not matter whether 70 or 100 GW of new capacity was built in a year—because within a few months there would be a further 50 GW of growth in demand—there was limited value in reforms trying to optimise decision-making, add validation or documentation burdens, or that risked slowing down effective progress. The cost of an electricity shortage is always higher than the cost of a similar level of excess supply, reflecting the fundamentally interdependent way that electricity affects almost every aspect of life. But when electricity demand growth slows to the point where it is no longer so difficult to keep up, it is time to consider introducing more nuanced approaches to influencing investment and operational decisions.

China is not alone in attempting to transition from a period of excess supply to a more balanced state. In the 1970s, an unexpected slowdown in growth in the United States—following decades of 7 per cent growth in electricity demand—left the economy with huge amounts of excess capacity. A change in fuel markets following the 1973 formation of OPEC and cost-overruns in the US nuclear power industry contributed to a triple whammy that resulted in regulatory consideration of whether costs had been incurred that consumers should not pay for. Efforts to reform the regulatory system and introduce market-based approaches also triggered consideration of stranded costs (costs that would not be recoverable under proposed arrangements, but which would have been recoverable under the prevailing regime). Changing the future need not require all legacy decisions be over-turned: cost recovery for legacy projects can be protected through transitional arrangements. Such backwards-looking stranded-cost recovery mechanisms enabled support for necessary forward-looking reforms.

Long before reaching these more nuanced considerations, China's governance arrangements—amongst the most complex and overlapping in the world—will need to be reconsidered. China appears to be moving towards more market-based mechanisms, with a strong reform agenda in place and many local initiatives underway. Such 'local' initiatives might easily span a province with an economy the size of several countries: small can be big in China. As noted above, China's most significant imbalances arise because of factors that cannot be managed wholly within a single province. Energy resources are not located where industry requires electricity, and this is particularly true of renewable energy resources. Any overall optimisation of resource and energy use requires national-level governance and regulation. The required reforms are massive. The United States offers some models in combining national with state-level energy regulation depending on whether the situation crosses a state border (and thus becomes subject to national regulation) or remains within a single state (in which case state regulation applies). Europe offers similar insights from the challenges of EU energy integration. But none of the available examples forms a perfect roadmap for success—they are more sources of insight into potential problems to be avoided.

Thus far China has managed to side-step many of the transition costs by internalising them through state-owned enterprises and through mergers (and rationalisations), again through state-owned enterprises. Ultimately, the financial bill is likely inescapable, and the only possible payers are tax-payers, consumers or investors. Exactly what mechanisms will be chosen to achieve the necessary compromises remains unclear even to the most experienced experts. The scale of China's energy imbalances is daunting.

And so, as China looks to move more towards market-based approaches for key aspects of its electricity sector, it should not surprise anyone to see such moves being developed with 'Chinese characteristics'. The fundamental challenges are familiar. The scale is what is different. The art of compromise—the sharing out of the pain of disruptive transition—is both an essential ingredient of success and hard-wired into the Chinese approach. That is a good thing. In many ways, achieving compromise has been the most consistently difficult challenge of energy reforms in the Western world.

This book has given a taster of the challenges and complexities in the world's largest electricity industry. The hard but intensely exciting work is just beginning.

INDEX

© The Author(s) 2018
L. Lester, M. Thomas (eds.), *China's Electricity Sector*,
https://doi.org/10.1007/978-981-10-8192-7